LETTERS ADDRESSED TO H.R.H. THE GRAND DUKE OF SAXE COBURG AND GOTHA ON THE THEORY OF PROBABILITY

LETTERS ADDRESSED TO H.R.H. THE GRAND DUKE OF SAXE COBURG AND GOTHA ON THE THEORY OF PROBABILITY

M.A. Quetelet
[Adolphe Jacques Lambert]

ARNO PRESS

A New York Times Company
New York • 1981

Editorial Supervision: RITA LAWN

Reprint Edition 1981 by Arno Press Inc.
Reprinted from a copy in the University of Michigan Library

THE DEVELOPMENT OF SCIENCE: Sources for the History of Science
ISBN for complete set: 0-405-13850-4
See last pages of this volume for titles.

Manufactured in the United States of America

Library of Congress Cataloging in Publication Data

Quételet, Lambert Adolphe Jacques, 1796-1874.
    Letters addresses to H.R.H., the Grand Duke of
Saxe Coburg and Gotha, on the theory of probability.

    (The Development of science)
    Reprint of the 1849 ed. published by C. & E. Layton,
London.
    1. Probabilities. 2. Statistics. I. Title.
II. Series: Development of science.
QA273.Q513 1981   519.2   80-2143
ISBN 0-405-13950-0

LETTERS

ADDRESSED TO H. R. H. THE GRAND DUKE OF SAXE COBURG AND GOTHA,

ON THE

THEORY OF PROBABILITIES,

AS APPLIED TO THE

MORAL AND POLITICAL SCIENCES.

BY

M. A. QUETELET,

ASTRONOMER ROYAL OF BELGIUM, CORRESPONDING MEMBER OF THE INSTITUTE
OF FRANCE, ETC. ETC.

TRANSLATED FROM THE FRENCH BY

OLINTHUS GREGORY DOWNES,

OF THE ECONOMIC LIFE ASSURANCE SOCIETY.

LONDON:
CHARLES & EDWIN LAYTON, 150, FLEET STREET.

MDCCCXLIX.

LONDON:
C. AND E. LAYTON, PRINTERS,
150, FLEET STREET.

TRANSLATOR'S PREFACE.

MANY are deterred from the study of the Theory of Probabilities from a mistaken notion that a previous acquaintance with the more abstruse parts of Mathematics is essential to the acquisition of such knowledge.

Though such may be the case in reference to the higher applications of the science, and for the purpose of mastering such works as those of POISSON and LAPLACE, still a large and interesting portion of the subject is susceptible of explanation on elementary principles; and M. QUETELET has moreover shown, in the work now presented, that a very useful portion of the subject may be conveyed in ordinary language.

The application of the Theory of Probabilities to Social, Political, and Moral Laws, is a matter of sufficient importance to engage the attention of the Statesman, the Philosopher, and the Jurist; and every attempt to simplify such knowledge conduces to the interest of society.

M. QUETELET has been eminently successful in this effort; for, by the clearness of his reasoning, the felicity cf his illustrations, and the perspicuity of his style, he has rendered the subject familiar to the ordinary reader.

With a view of drawing attention to this department of science, and of extending the utility of M. QUETELET's work, I have been induced to offer this translation of it to the English public.

The French weights and measures have been converted into English in all cases where it was requisite, or better adapted to convey the Author's meaning.

It was my intention to have so enlarged the notes as to have furnished a mathematical demonstration of every principle enunciated in the work; but I have been compelled by ill health to abandon that design.

The work is therefore given in its original form, with no other alterations than those above mentioned, and a few corrections of the errors of the press.

O. G. D.

INTRODUCTION.

CERTAIN circumstances, which have left me many pleasant reminiscences, made it necessary for me (nearly ten years since) to devote my whole attention to the application of the Theory of Probabilities to the study of the moral and political sciences.

I then felt how desirable it was that this science should be made more *elementary*,—and that it should be brought down from the high regions of analysis, and placed within the reach of those who are most often obliged to make use of it. It links itself in reality to a great number of questions which interest both the legislator and the man called to the management of public affairs: both are often under the necessity of reading statistics of the past, and of endeavouring to collect from that source whatever may be useful for the future; they feel the want of means to appreciate the results produced by modifications introduced into the laws, and, in a certain measure, to ascertain the weight that should be attached to symptoms which announce the adversity or prosperity of the country.

This work, which was begun in 1837, has been written in the form of letters, in reply to a request as honourable as it was flattering to me. However, motives of convenience led me some time after to suspend a correspondence to which I attached the highest value; and I should have thought it my duty to abandon it altogether, if the august Princes to whom it was addressed had not deigned to request me to complete my work, and to authorize me to publish it under their auspices.

I am fully sensible of all the obligations which such a mark of favour imposes upon me; and I have exerted myself to justify it. It was, besides, well fitted to recompense me for the difficulties I have met with in seeking to make easy of access many details which no one had ever attempted to translate into ordinary language. I may mention in particular all which relates to the effects of accidental causes in the appreciation of dimensions.

Not content with having shown the principles of the Theory, I have endeavoured to demonstrate them by Elementary Algebra; but I have thought right to transfer this part to the notes to the work. I had already published some of the principal results in the second volume of the *Bulletin de la Commission Centrale de Statistique de Belgique*, as a first step in the path I proposed to pursue.

As my work developed itself, I perceived the necessity of separating it into two parts,—the one relating to the application of the Theory of Probabilities to moral and political sciences; the other having for its object the study of the laws which relate to the development of man and of the body social.

This last work (to be produced under the auspices of H. R. H. Prince Albert) will form a continuation of the work I published in Paris in 1834, entitled *Sur l'Homme et le Développement de ses Facultés, ou Essai de Physique Sociale.*

I have deemed it right to suppress dates generally, since the greater part of the letters have been modified, and do not follow the order in which they were in the first instance written.

I do not disguise from myself the many difficulties which attach to the employment of mathematical methods in the analysis of phenomena, and especially moral phenomena. This study, as yet new, has I know alarmed many readers, who have fancied they have seen in it a tendency to materialize that which belongs to the most noble faculties of man. I will show hereafter how ill-founded is this charge.

In the work which I now present, I examine more especially the course to be pursued in the study of purely material phenomena. I have given particular attention to statistics,—a science which is far from being understood, although its utility is generally recognised in proportion as it is cultivated with discernment. I am far from pretending to give a treatise on this subject: I have simply limited myself to pointing out many errors into which authors frequently fall, and to throwing out some ideas which have appeared to me to cast light on the Theory.

BRUSSELS, 18*th December* 1845.

TABLE OF CONTENTS.

PART THE FIRST.

ON THE THEORY OF PROBABILITIES.

PART THE SECOND.

ON MEANS AND LIMITS.

PART THE THIRD.

ON THE STUDY OF CAUSES.

PART THE FOURTH.

ON STATISTICS.

NOTES.

LETTERS

ADDRESSED TO H. R. H. THE GRAND DUKE OF SAXE COBURG AND GOTHA,

ON THE

THEORY OF PROBABILITIES,

AS APPLIED TO THE

MORAL AND POLITICAL SCIENCES.

FIRST PART.

ON THE THEORY OF PROBABILITIES.

LETTER I.

OUR KNOWLEDGE AND OUR JUDGMENTS ARE IN GENERAL ONLY FOUNDED ON PRO-
BABILITIES MORE OR LESS GREAT, WHICH WE SHOULD KNOW HOW TO VALUE.

BRUSSELS, 30th *April*, 1837.

THE interest which your Highness has seemed to take in the
study of the Theory of Probabilities, and especially in the application that can be made of it to the moral and political sciences,
encourages me to hope that the new developments which I shall
have the honour of presenting to you will not be unworthy of your
attention. There is scarcely in fact a branch of our knowledge
whose aim can be more philosophical, or more directly useful.

From its origin the Theory of Probabilities has been cultivated
by the most distinguished minds, by the most profound thinkers:
it has ranked among its promoters Pascal, Fermat, Leibnitz,
Huygens, Halley, Buffon, the Bernoullis, D'Alembert, Condorcet,

B

Laplace, Fourier, and it may be said the greater part of the learned men who have the most powerfully influenced the age in which they have lived. It has equally attracted the attention of many statesmen of eminence, who have known how to appreciate the fruitful results which they had a right to expect from it. However, this theory, so valued by the finest geniuses of modern times,—this theory, which ought to be the base of all sciences of observation, is not only untaught in the schools, where so many things (and one might almost say so many useless things) are taught, but it is scarcely known even by the men who have the greatest interest in its use. These would be grounds for astonishment, did not history teach us how long a time is necessary for the most beautiful discoveries to descend and make their way to the masses, where they are destined to become fixed and to produce their finest fruits.

The sciences make so much the more rapid progress as that acquired knowledge becomes more exact, and the means of expressing it more precise. But we are so little advanced in this respect, and above all in the sciences of observation, that we every instant confound certainty with probability, and that which is probable with that which is but possible.

These mistakes are not only met with among men of the world, but many remarkable instances among learned men of otherwise incontestible merit might be cited. I have seen more than one historian revolt at the idea of ranging the suicide of Demosthenes amongst the number of probabilities; or of not regarding it as certain that Aristotle drowned himself in the Euripus, overwhelmed with grief from his inability to discover the cause of the ebb and flow of the waters of that river. I have seen other learned men repel with disdain the assertion, that many of the facts with which we are acquainted in Astronomy, or even in Physics, only rest on probabilities more or less great. It is, however, incontestible that *certainty* exists in but very few things,—mathematical truths, for instance among the number; but this does not tend to establish that we should give ourselves up to scepticism, and revoke in doubt the benefits of science.

The principle of attraction has enabled us to explain with an admirable simplicity the phenomena of our planetary system. It has made the appreciation of the different movements which observation had recognized in the celestial bodies to depend on the solution of a problem in analytical mechanics. Much more,—this same principle, subjected to calculation, has furnished evidence of facts which would have escaped the sagacity of the most experienced observer, armed with instruments the most delicate. Who will however dare to pretend that this principle is the most complete expression of a law of nature? Who can affirm that this is not a particular case of a much more general law? or that the results they have deduced from it are not values sufficiently approximative, since the neglected quantities are not appreciable in the present state of science? Notwithstanding the reasons we have for not attributing certainty to the principle of attraction under its actual enunciation, we are nevertheless able to make use of it with the greatest success in calculating the phenomena of our planetary system, and in predicting their return with a precision which is the object of man's general admiration.

The same might be said of the greater part of the principles of Physics, and of all the sciences which rest on observation. We have not sufficient reasons to accept them as certain, and as if they announced in all its plenitude nature's rule of action; but the probabilities which we have for considering them such are in general so strong, that we have no difficulty in attributing to them in most cases a value equal to certainty.

There is an infinity of intermedials between impossibility and certainty. To pass from one to the other, the probability may vary by an infinite number of gradations; and among a hundred assertions there are not perhaps two which represent the same degree of probability. These gradations we nearly always ill appreciate, and consequently our opinions are more or less inaccurate. To many persons events are scarcely probable, except in one way: if the probability of an event is very great, they take it for certainty; if, on the contrary, it is very small, they estimate the event as impossible.

The sciences could not be satisfied with such defective estimates. These vague perceptions are at most tolerable in the ordinary affairs of life. Thus, I look upon it as certain that I shall finish this sentence without leaving my seat; yet, even before my pen can reach the end, an unexpected visit may suddenly interrupt me, —I may meet with some accident,—I may be suddenly called from the room by the mournful cries of some one in the house who may have been hurt by a fall. All these accidents, and many others impossible of enumeration, have in my eyes so small a probability of happening that I can well disregard them. Further, continuing to use it as the men of the world do, I have no difficulty in looking upon it as certain that this letter will be finished, and forwarded to your Highness by to-morrow at the latest; and yet here other unforeseen events may be very numerous. If even I can avoid them all, and attain exactness on my part, will it be the same with those who will be charged to carry the letter, or to take care that it is sent? As to knowing if the letter will reach its destination, I look upon its remittance as probable; I consider it quite as probable that your Highness will not only receive my letter, but that you will read the whole of it. Nevertheless, notwithstanding what my vanity (as the writer) may tell me, this last circumstance has but few chances in its favour.

It would be interesting to find out what are the approximate limits probability ought to attain, in order that it might assume in the eyes of the men of the world the degrees of certainty or impossibility. These limits are not the same for all men, nor with the same person for all things indiscriminately. A variety of circumstances influence our judgment in this respect: our passions especially often place in the same rank events whose probabilities are essentially different. The variations of probability are to the understanding what the gradations of colour are to the practised eye, or the diatonic scale to the ear that can appreciate all its degrees.

A great number of questions connect themselves with the Theory of Probabilities of a nature calculated to excite the curiosity of, and which deserve the greatest attention from, both the philosopher and the statesman. This your Highness already well understands; but you may be better convinced by what follows. Among the many different applications which can be found of this Theory, I shall give the preference to those which relate to man in his social state : we shall find there a fruitful mine of useful discoveries, of which our predecessors have scarcely caught a glimpse, and which still remains almost wholly unexplored.

The computation of probabilities is but the instrument which should regulate the labour of working matters out; but it becomes indispensable in the researches to which we wish to apply ourselves. It ought, in fact, to enable us to distribute with advantage the series of our observations,—to estimate the value of the documents that we may use,—to distinguish those which exercise the greatest influence,—and afterwards to combine them so that they shall err the least possible from truth, and to calculate definitely the degree of confidence to be placed in the results obtained. The Theory of Probabilities only teaches us in the main to do with more regularity and precision that which even the most judicious have hitherto done in a manner more or less vague. It tends, moreover, in the phenomena with which we shall have to occupy ourselves, to substitute science for that which is conventionally called practice or experience, and which is most frequently but a blind routine.

In this respect also, prejudice is so deeply rooted, and prepossession is such that every instant the most strange assertions may be expected from the mouths of otherwise skilful persons. It is now a well-proved fact, although the cause is unknown, that there are generally more boys born than girls. Well, announce this fact in the presence of an accoucheur who is not aware of it, he will no doubt tell you that his experience has shown a contrary result. Then ask how many observations his experience comprehends, he will answer you, without exposing himself to a charge of exaggeration, that he could quote more than a thousand,—what do I say? —more than two thousand,—more than three thousand. Ask again if he has taken the trouble to register all these observations, and he will immediately appeal to his memory. You will then see that these two or three thousand observations which he advanced reduce themselves simply to those which most particularly struck him, and which will have contributed to form what he designated his experience.

Indeed the opinions which we form for ourselves most frequently rest on bases equally deficient in solidity. What science had a right to demand from the practitioner to whom I have just alluded, is, that he should have carefully registered the facts, and not have confided them to his memory,—that he should have made a *complete enumeration*, and not have contented himself with citing isolated facts which had more especially fixed his attention, because they were perhaps more in accordance with his prejudices. Lastly, supposing even that the indications had been taken with the greatest care, but without systematic arrangement, and to the number of three thousand, there would not have been a sufficient number to establish an opinion of even a tolerable degree of exact probability. The number of male births, in fact, exceeds but slightly the number of female births. The ratio is 106 to 100 nearly, for the whole of Europe. When a large population is operated on, that of a kingdom for instance, this ratio is nearly constant every year; but it may oscillate between very wide limits when it refers to a populous town, where it is not uncommon to notice during a year more girls born than boys.

Thus our practitioner might, even were the assertion he made correct, not have weakened the fact generally observed. His error would only have been that he had drawn conclusions from numbers too small to establish the fact he wished to prove. His result had not the degree of probability necessary for its acceptance with confidence.

The greater part of the errors and the differences which are found in works of statistics have no other source; and the world does much wrong, in such cases, in blaming the science rather than the calculators, who alone ought to be responsible.

So as not to enter on a wrong path, it will be prudent to repeat summarily the principles of the Theory of Probabilities, before devoting ourselves to researches on the manner of procedure in the sciences of observation, and in statistics in particular. This will be the object of my next letters.

LETTER II.

BRUSSELS, 29*th May*, 1837.

I AM happy to learn, by the letter that your Highness has done me the honour to address to me, that you appreciate the utility, and I will dare to say the pleasure, which attaches to the kind of studies which are about to occupy us. If anything, however, could have pained me in reading your letter, it was to find that your Highness had been indisposed since your arrival at Bonn.

I hope, however, that the details on which I would fix your attention will not be too dry for a convalescent.

The principles of sciences rarely present attraction; but if the entrance to the edifice is not the most brilliant part, it ought at least to be easy of access and conveniently lighted. I will ask, then, permission to remind you of certain principles of the Theory of Probabilities, in order that we may understand one another as to the value of many words which will be hereafter of frequent recurrence.

When there are many ways in which an event may occur, they are called the *chances* of this event. Thus the drawing of a card from an ordinary pack presents 52 chances, since any one of the 52 cards of which the pack is composed may be taken indifferently.

When the nature of the event which is wished for is defined, there exist two kinds of chances, the one *favourable*, and the other *unfavourable* to the wished-for event. Thus a person desiring to draw a picture-card from a pack of 52 cards would have 12 chances in his favour, and 40 chances against him.

If all the chances are favourable, their total is certainty.

An event is arbitrarily said to be *probable* when there are few chances unfavourable to its happening, and to be only *possible*, or but little probable, when the number of unfavourable greatly exceeds the number of favourable chances.

It can be conceived that all events are not equally probable. *The mathematical probability is estimated by the number of chances favourable to an event divided by the total number of chances.* According to this principle, the probability of drawing a picture-card from a pack of 52 would be the fraction $\frac{12}{52}$: there are, in fact, 12 favourable chances out of the whole number of 52 chances.

The probability of the wished-for event *not happening* is estimated in the same manner,—that is to say, by dividing the number of unfavourable chances by the total number of chances. Thus in the preceding example the probability of the event not happening is $\frac{40}{52}$.

In general, every uncertain event gives rise to two opposite probabilities; namely, that the event will happen, and that the event will not: *the sum of these two ought to be equal to unity.* Unity becomes thus the symbol of certainty.

The application of the Theory of Probabilities would present but few difficulties, if all the different possible chances could be enumerated, and if all the chances were rigorously the same. But it is not so; and in certain cases much sagacity is required to guard against great errors in these kind of appreciations. Of all games, the most simple is without dispute that of "pitch and toss," since there are but two possible chances, and the coin must necessarily fall on one or other of its two faces. The probability of each of these events is then $\frac{1}{2}$. However, if the coin were not homogeneous, if it were constructed in such a manner that it would more easily fall on one face than the other, it is evident that the two probabilities could not be respectively represented by the fraction $\frac{1}{2}$. If the coin, for example, fell regularly twice on one face, while it fell but once on the other, the tossing of the coin might be considered as presenting three possible chances, two of them in favour of one face, and one of the other: the respective probabilities would be $\frac{2}{3}$ and $\frac{1}{3}$.

Fraudulent players sometimes contrive a like advantage in playing with loaded dice. Where a die of six faces is well made, it ought to fall with equal facility on each of its faces: the throwing of the ace, for instance, ought to have the same probability as the throwing of any other number. However, the die can be so altered, either with respect to the homogeneity of the mass, or with respect to the shape, as to render the chance of throwing the ace as great or as small as may be wished.

It can be understood, that a game could be played with such a die, provided an exact account could be kept of the inequality of the chances of each face turning up.

The difficulty which we here meet with presents itself every instant in the appreciation of the probabilities which relate to the phenomena of nature. The difficulty becomes greater still in case the die (to continue the comparison) not only presents unequal faces, but even the number of faces be unknown.

Such is in effect the condition in which we place nature when we seek to sound its secrets, and to value the respective probabilities of events which may occur. One often thinks to have foreseen everything, and to have carefully enumerated the circumstances which could present themselves; and afterwards with astonishment finds that the event, when it has happened, is not any of those which were expected. It is then said that chance brought it about. But what does this word mean, except it be our ignorance that our die had another face, which we had not perceived, and on which we did not suppose it could fall? The word *chance* serves conveniently to veil our ignorance: we employ it to explain effects of whose causes we are ignorant. To one who knew how to foresee all things there would be no chance; and the events which now appear to us most extraordinary would have their natural and necessary causes, in the same manner as do the events which seem most common with us.

Another comparison, taken also from the theory of games of chance, may give some further light. Suppose an urn were presented filled with balls which only differ in colour, and the probability is demanded that the first ball drawn will be a white one. It is evident that, in order to form a judgment for which we have some motive, we must have some preliminary information. To make this information complete, we will empty the urn, and examine what number of balls of each colour is contained therein. I suppose, then, that there are in it two white, three black, and four red balls, in all nine balls: we should then say that the probability of the first ball drawn being white is $\frac{2}{9}$.

This estimate it is seen presents no difficulty, when we can be assured of the number that the urn contains, and of the manner in which they are distributed as regards colour. In general, in the different games which we call "games of chance," the number of chances is limited, and their nature known; but it is not the same with what relates to the sciences of observation. The urn is open before us,—we are allowed to draw from it as often as we will, to multiply proofs at leisure: but this urn is inexhaustible, and it is only by induction that we can know what it contains.

We are reduced then to set out from new considerations, to estimate the probability of an event when the number of chances has no limit, and when we are ignorant how the chances are distributed. This inconvenience unfortunately presents itself in most of the cases which will occupy our attention,—that is to say, in the appreciation of the probabilities of social and natural phenomena. This appreciation is of sufficient importance to form the subject of a separate letter; and it presents so much interest that I presume your Highness will not desire to see deferred the explanations into which I shall be forced to enter.

LETTER III.

ON THE PROBABILITY THAT AN EVENT OBSERVED SEVERAL TIMES IN SUCCESSION
WILL OCCUR AGAIN.

OSTEND, 25*th July*, 1837.

I AVAIL myself of the first moments of leisure that I have found
at Ostend, to renew a correspondence which I deeply regret has
been interrupted. The place from whence I write is dear to your
Highness, and I should be happy if my letter could lessen its
distance.

At least, not to stray too far from this illusion, I pray your
Highness to transport yourself in thought to the shores of the sea.
It is there that we will to-day hold our peaceful conference, on the
verge of this immense ocean, the limits of which the eye can no
more comprehend than human intelligence can compass the boundaries of the vast field of science.

We are here then before these foaming waves, which seem to
invade in their triumphant march the gently-sloping beach, which
they had but a few instants before abandoned. Their white and
roaring lines closely follow each other, and unroll themselves majestically, like the close ranks of an army, which advances in good
order, and shouts its war-cry as it dashes on the foe. They
already strike with vigour, and cover with their foam, these long
jetties and palisades, which human industry has placed far in the
sea, like advanced sentinels, to protect the progress of the vessels.

This spectacle is sublime, but we must learn to understand it.
It frightened, they say, the barbarians who in ancient days ravaged
our country, and strove to combat with the sword this new enemy
which seemed to wish to dispute their conquest. However absurd
this act of folly appears to us, we can yet explain it. In effect,
if we had no idea of the phenomenon of tides, and if placed before

the ocean we were suddenly to perceive the invasion of its waters, we might retreat in fright, and we might fear lest this progressive invasion, growing nearer and more near, might swallow us together with the soil on which we stood : we should undoubtedly look on the event as a prodigy.

But the retreat of the waters, which would soon after take place, would afford us hope; and when the sea rose, at the succeeding tide, we should no more feel our former fear. At the end of a few days even, struck with the regularity of the phenomenon, we should conclude by ranking among ordinary events that which we had at first considered a prodigy. Still further, we should be much astonished not to see, after the retreat of the waters, that rising movement re-occur which was the cause of our fear at first : we should begin already to look upon the re-occurrence as probable, without even knowing whether its reproduction be the result of constant causes or be purely accidental.

It is nearly the same with a number of events which in former days caused the astonishment and the fear of man; such as eclipses, the appearance of comets, and other phenomena, of which in ages of ignorance impostors so often availed themselves.

If we have recourse to science, it gives us the following rule to estimate the probability of the return of an event which has reproduced itself periodically several times in succession. *Divide the number of times the event has been observed increased by unity by the same number increased by two.* Thus, after having seen the sea rise periodically ten successive times at an interval of about twelve hours and a half, the probability that it will rise again for the eleventh time would be $\frac{10+1}{10+2} = \frac{11}{12}$.

Now that we have seen the tide after a series of ages re-occur on our coasts in the most regular manner, we may regard at the moment of low tide the rising movement of the waters which is about to follow as presenting a probability very nearly equal to certainty. So that, contrarily to what we thought at first, we should consider it extraordinary or prodigious if the sea did not rise at the hour of the tide.

We should consider it equally extraordinary if the sun, which is now shining with so much majesty above the waters, did not gradually sink, to disappear after a time below the level of the seas. The habit of beholding this spectacle, and our astronomical knowledge, have long since familiarized us with this phenomenon, which undoubtedly would appear inexplicable were we to witness it for the first time, and without being prepared for it. The ancients were far from being able to account for it. Many Greek philosophers had the strangest ideas of its nature. Some regarded the sun as an inflamed mass, which plunged itself every night into the waters of the sea; and they pretended to have heard a hissing noise like that produced by plunging a red hot iron into water.

Whatever may have been the difficulty of explaining the setting of the sun, it is certain that there was no one who did not conclude by believing the necessity of its periodical return, and by even regarding the arrival of this event as a certainty.

As for ourselves however, whose knowledge is more advanced, and who can explain the phenomenon in the most simple and satisfactory manner, we only regard its return as a probability. Whence has this apparent contradiction birth? Your Highness has already seen that mankind in general do not admit of graduation in the degrees of probability of the events which pass around them. A thing can only appear to their eyes as very doubtful or certain. However, we have but probabilities (extremely great, 'tis true,) to believe that the natural laws, which we see manifest themselves with so much regularity, will still manifest themselves in the same manner in the sequel. We are ignorant, for example, whether our planet, by the shock of a comet or by other causes, may not be subjected some day to constantly present the same face towards the sun, in the same manner as the moon does with respect to the earth, which would produce in one of our hemispheres perpetual day, and in the other perpetual night. Already, in our polar regions, we can form an idea of such an order of things, although it is difficult to foresee the general disorganization which would manifest itself among animals and plants. I leave to the imagination the care of putting everything in the place best suited for it, and will return to more positive considerations.

From what we have already seen, the method of calculating the probability of sunset, not taking into consideration the other scientific motives we now have for believing in its return, is as follows. On the 1st January, 1837, 5,841 years of 365 days, or 2,131,965 sunsets, had been counted since the creation, which we carry back to 4,004 years before the Christian era. The probability that another return will take place at this epoch was then $\frac{2131966}{2131967}$, —that is to say, a fraction approaching so nearly to unity that it may be considered equivalent to certainty. It must be remarked, that the happening of the second sunset presents a smaller probability, and that the probabilities of sunsets become so much the smaller as the times for which they are calculated are more distant. The fraction which expresses the probability of a distant sunset may vary so sensibly as to cease to have the value of certainty even in the eyes of the community.

In the sciences of observation, and in particular in the application of the Theory of Probabilities to social events, it is very important to remark that an intimate knowledge of the past may lead to pronouncing almost with certainty what will *soon* happen; but the deductions that we make become so much the less probable as the events we foresee extend the farther into futurity. Thus, to calculate the probability that an event observed any number of times in succession will re-occur, this rule should be followed. *The probability is equal in value to a fraction which has for its numerator the number of observations plus 1, and for the denominator the same number plus 1, and also plus the number of times that the event is to re-occur.* Thus, on the 1st January, 1837, the probability of the occurrence of five successive sunsets was $\frac{2131966}{2131971}$.

To avail myself of a comparison which the illustrious Laplace made use of in a circumstance different to this, we can assimilate what is passing in the same manner that we can see things by the interposition of one or more glasses, which will gradually extinguish the clearness of the object according as what we foresee is more distant from us.

LETTER IV.

ON THE PROBABILITY THAT AN EVENT OBSERVED SEVERAL TIMES IN SUCCESSION
DEPENDS ON A CAUSE WHICH FACILITATES ITS REPRODUCTION.

BRUSSELS, 25th October, 1837.

IN my last letter I had the honour of explaining to your Highness how, after having observed several reproductions of the same
event, to calculate the probability that this event will manifest itself
again one or more times. We had occasion to remark that the
probability acquires importance in proportion as the number of
times that the event has been observed is greater, and as our
inquiries extend less into the future.

A second question naturally presents itself here: it is, whether
the event which has been seen to occur several times in succession depends solely on one cause, or on the combination of a
certain number of causes; or, again, whether it be purely accidental. Thus, to return to the example I have chosen, we ought
to have inquired if there existed one cause, or a combination of
causes, in favour of the periodical return of tides. And this question relative to the origin of the phenomenon ought to have been
asked nearly at the same time as that which related to its next
reproduction. This question then, so natural to the curiosity of
man, and to which good sense would already have found an approximate answer, has already been resolved by the Theory of
Probabilities.

We were sensible that the more frequently the event was manifested under the same circumstances, the more probable it became
that it was the result of one sole cause, or of several simultaneous
causes; but this probability presented itself under a vague form.
The English geometrician Bayes proposed the following rule to
appreciate its value. *When the same event has been observed several*

times in succession, the probability that there exists a cause which facilitates its reproduction is expressed by a fraction which has for its denominator the number 2 multiplied by itself as many times as the event has been observed, and for its numerator the same product minus 1.

After having seen the sea rise periodically ten times in succession, at intervals of about $12\frac{1}{2}$ hours, if the probability that it would rise the eleventh time were required, we should have, as I have already said, $\frac{11}{12}$. According to the preceding principle, the probability that there exists a cause which necessitates the reproduction of this phenomenon will be $\frac{2047}{2048}$.

Your Highness will see that we have more reason to believe in the existence of a cause which has facilitated ten times in succession the reproduction of the same phenomena, under the same circumstance, than in its reproduction for the eleventh time. This distinction science alone permits us to establish; and simple good sense, however great its extent, would never have arrived at it, even approximately.

In general, the probability that there exists a cause which necessitates the reproduction of an event increases much more rapidly than the probability of the next occurrence of the event. This results from the formulæ themselves, which serve to calculate the respective probabilities.

I had occasion to observe in 1835, in the night between the 10th and 11th of August, a great number of falling stars: it had been the same in 1834. These observations were registered, and I lost sight of them until towards the close of last year; at which time I returned to the extraordinary appearances of falling stars, with which I had occupied myself much in my youth. I could then prove that the same phenomenon, according to different observers, had re-occurred several times at the same period of the year. From that time I thought myself right, not only in concluding that there existed very probably a cause which facilitated this return, but also in announcing the probability of a next return. This, in fact, I did; and the event confirmed my conjectures. I am now able to prove, since the commencement of the present century, eighteen periodical returns towards the middle of August; and

I have reason to believe that this spectacle, which moreover does not appear to be visible simultaneously at all points of the globe, may have escaped attention so many years during the present century either from want of observers or from the obstacle of cloudy nights. Has the cause which has given birth to this phenomenon only manifested itself in these latter days? and will it continue indefinitely in action? Will it act intermittently, and at intervals more or less long? There may be falling stars as there are ordinary showers. A succession of bad days does not increase the probability of having other bad ones. Experience teaches us in fact that rain has longer or shorter intermissions, and that, when prolonged to excess, the probability of its continuance, far from augmenting, diminishes.

LETTER V.

I HAVE hitherto supposed that the expected event only manifests itself in one manner; but your Highness will see that this case is in some degree exceptionable.

There will in general be presented two kinds of events; the one favourable, and the other unfavourable to the trial. In this case theory offers the following method of calculating the probability that one of two species of events will occur once again. *Divide the number of times the event has been observed to happen plus 1 by the total number of observations plus 2.*

I will give an example of the calculation. From the observations made down to a recent date, it was believed to have been recognised that, not only the movements of translation, but also the movements of rotation of the planets and their satellites were accomplished in the same manner. " A phenomenon so remarkable," says Laplace, in his *Philosophical Essay on Probabilities,* " is not the effect of chance; it indicates a general cause which has determined all the movements. To ascertain the probability with which this cause is indicated, we shall observe that the planetary system, as known at the present day, is composed of eleven planets and eighteen satellites (at least if with Herschel we attribute six satellites to the planet Uranus). Rotatory motion has been recognised in the sun, in six planets, in the moon, the satellites of Jupiter, the ring of Saturn, and one of his satellites. These motions with those of revolution form a total of 43 movements in the same direction. We find then, by the analysis of the probabilities, that there are four million million chances to one that this disposition is not the effect of chance, which forms a probability much superior to that of historical events which we are not permitted to doubt."

We will add, that with the preceding data, and in case of the discovery of a new planet, the probability that this new planet will have a motion of translation with the same direction as the others should be represented by the fraction $\frac{44}{45}$.

But it is found in the present day that, after having submitted Uranus and its satellites to new observations of a very delicate kind, which could only be done with the most powerful instruments, Sir John Herschel, son of the celebrated astronomer to whom the discovery of this planet is owing, has been able to recognise, as his father had announced, that two of the satellites, whose orbits make large angles with the ecliptic, effect their movement of translation not from west to east, but in the contrary direction. This discovery must necessarily modify the results of the calculations I have cited above. We have in effect, of 43 motions, whether of translation or rotation, observed up to the present day, not 43 but 41 direct motions, and two retrograde, which gives us the probability that, if another planet or a satellite were discovered, the motion would be direct, the fraction $\frac{42}{45}$, and for the contrary probability $\frac{3}{45}$.

When several species of possible events exist, and it is wished to calculate from previous observations the probability of each of them, the calculations become considerably complicated. Happily the appreciation can be reduced to a simple principle, and the deviation from rigorous exactness is but small when the observations are numerous. This principle is, that *the favourable and unfavourable chances may be considered as being numerically in the same ratio as the observed events to which they refer.*

I will illustrate this by an example. Suppose an urn presented containing a considerable number of balls of different colours, and the probability is required that the first ball drawn will be white. It is evident that, to answer this question, the number of balls the urn contains, and their colour, should be known. The problem then would resolve itself *à priori*, as I have already shown in one of my former letters. But not having this knowledge, to obtain a first idea of the contents of the urn, I extract some balls, and I replace them in the urn, after each drawing, in order that the same conditions may exist. If, after a certain number of drawings, I have only brought out black and white balls, I may consider that the urn contains only balls of these two colours. If I continue the drawings during a whole day, and uniformly obtain the same results, my judgment will receive so much more support.

Let us admit, for an instant, that the number of white balls withdrawn is sensibly equal to the number of black ones. I should have reasons to believe that the reality is in conformity with the results of experiment, and that the black and white balls in the urn are in equal proportions. If, on the contrary, the number of white balls predominated, I should in like manner be informed of it by the successive trials I should have made. The ratio of the number of white balls to the number of black balls, after a great number of drawings, would approximate the more to the ratio actually existing in the urn, in proportion as the trials were repeated. I can then find out, with as great a degree of precision as I wish, by sufficiently multiplying the trials, first, that the urn only contains black and white balls, and secondly the numerical ratio which exists between each colour.

The urn, then, which we interrogate is Nature. We can multiply our experiments to infinity; we need not even use the precaution of replacing the balls in the urn, for whatever we draw out does not alter the proportions of the remainder,—it is less than a drop of water cast into the ocean.

I am asked, if there be more boys born than girls. To procure the means of reply, and to return to the example previously chosen, I can compare each birth to the drawing of a ball from an urn of the contents of which I am ignorant. After a sufficient number of observations, I count the boys and girls. Supposing the trial to have been made during the year 1841, and that in all the rural districts of Belgium we had counted 53,437 male and 49,788 female births. The first number is greater than the second. Must it be concluded that this is in consequence of a law of nature which facilitates the birth of males, or that the excess of one of these numbers over the other is but accidental?

In order to obtain more light in this respect, recourse must be had to one or more years. The ratio of the two preceding numbers is expressed by 1·072,—that is to say, that this number of male births was counted to 1 female birth. But the ratio calculated for nine years, from 1834 to 1842, presents nearly the same number, that is 1·063. There is then truly a predominance in favour of male births, and the ratio nearly coincides with that which we have found by the numbers from 1834 to 1842.

One more example (and I will borrow it from Meteorology) will make the application of our principle better understood. Suppose it be required to ascertain whether the indications of the barometer are of any value in predicting the weather; and that, of 1,175 observations, we should have found 758 instances of the barometer being low at the approach of rain, and 417 in which it had made an upward movement. We might conclude that, according to experience, there might be reckoned (all other things being equal) 758 chances to 1,175 in favour of the fall of the barometer at the approach of rain, which gives to this event the probability $\frac{758}{1175}$. We should have $\frac{759}{1177}$ by calculating according to the method indicated at the commencement of this letter. These two values differ a little from one another: they have been deduced from observations made by Poleni in the environs of Padua. Van Swinden, in making analogous researches at Franeker, in the year 1778, found that on the approach of rain the barometer rose almost exactly as many times as it fell; so that this instrument would not be of any absolute value for this sort of appreciations.

These discordant results will explain themselves, if we consider the infinite number of different causes which act upon the barometer, and which would render necessary much more numerous observations than those hitherto collected. In order that the results might be rigorously comparable, they should have been obtained under the same atmospheric circumstances, and in the same climate. These kinds of appreciations are very delicate.

It is of the greatest importance to consider, before making like comparisons, whether the urns which are employed in the experiments (to return to our first example) are identically the same as to the manner in which the balls of different colours are distributed in them.

LETTER VI.

ON THE PROBABILITY OF A COMPOUND EVENT.

On reading again the conclusion of my last letter, I asked myself the question, whether it would not have been better to have passed over in silence the example relative to the indications of the barometer. This example is perhaps badly chosen; but it will at least possess the advantage of suggesting to us some useful results.

The barometric variation rarely depends on one single cause; it is in general a complex event, the product of a great number of causes,—and with a fall of rain it is the same. The causes, moreover, may vary by an infinite number of gradations, and may act either singly or in different ways in combination one with another: in fine, all those which act on the barometer do not act equally in inducing rain. Regard must therefore be had to the probabilities of each of them, when they concur in producing the expected event.

There are very few events which are not compound. Our least designs suppose a combination of causes, which should all be favourable to ensure success to our attempt. We should decline to form a project, especially if it related to some future time, if we were to calculate how many probabilities there were of our being deceived.

I might intend, in the course of this summer, to pay your Highness a visit; but I dare not promise myself this pleasure,—it is so subordinate to different causes, and the probability that all will be favourable is so small. In the first place, it is not certain that six months hence I shall be still existing, or that I shall be in sufficiently good health to undertake a journey. Supposing even that my health were very good, extraordinary duties, family reasons, or frightful times, might derange my projects.   You, Monseigneur, might be absent at the time at which I intended to make the journey, or even (notwithstanding your youth and the advantages of robust health) might be unwell, and little disposed to receive me. There yet remains an infinity of other causes, which I should have to enumerate; and so as not to err in my calculations, I ought to estimate the probability of each of them.   Let us see how these calculations are established.

When an event is *compound* (that is, when its occurrence depends on many causes independent of each other) the probability is calculated in a very easy manner.   "*Take separately the simple probability of each of the causes which influence the expected event, and multiply all these probabilities together,—the product will express the probability of the compound event.*"

An example will enable us to make clear what theory may leave doubtful in this respect.   I propose presently to take a walk. "What is the probability," I inquire, "that the two first persons I meet will be two men, supposing that there are in the streets as many women as men?"   The probability that I shall first meet a man is $\frac{1}{2}$,—that I shall meet a man again is also $\frac{1}{2}$; and the probability that these two simple events will occur successively, so as to produce the compound event which I look for, is $\frac{1}{4}$, the product of the two former fractions.   I have in fact but one of four chances, all equally possible, if we consider that the meetings might happen as follows,—

A man, and then a man.

A man, and then a woman.

A woman, and then a man.

A woman, and then a woman.

What is the probability of throwing an ace twice in succession with a die of six faces?   The desired result is evidently composed of two simple events,—to throw an ace the first time, and then to throw an ace a second time.   The probability of the first event is $\frac{1}{6}$,—the probability of the second is the same; and these two fractions multiplied together give $\frac{1}{36}$, the required probability.

Let us take another example in the sciences of observation. What is the probability that, if a register of the civil estate were opened, the first three births inscribed therein would be male births? The looked-for event is here composed of three simple events, which have each the same probability, which is about $\frac{106}{206}$. We know, in fact, that where there are 100 girls born, there are nearly 106 boys born. We must therefore make a product in which the preceding fraction will enter three times as a factor, and we shall have 0·136. The probability sought is very small, and would have been still smaller if we had calculated on the hypothesis of three female births taking place in succession. The simple probability of a female birth being $\frac{100}{206}$, that of three successive ones would be 0·114.

When the event is composed of a great number of simple events the probability decreases very rapidly. It may even be so small as to make it a matter of difficulty to form an idea of its value. I will give an example. Suppose it to be a question of taking, two hundred times in succession, a white ball from an urn containing equal numbers of white and black, and the precaution also taken of replacing each time the ball drawn, in order that the conditions may remain the same. We must here make a product in which the fraction $\frac{1}{2}$, the probability of drawing one white ball, shall enter two hundred times as a factor. This product then, which expresses the probability of the required event, is a fraction with unity for its numerator, and for its denominator a number expressed by 61 figures. In other words, there would be but one chance out of a number expressed by 61 figures. It would be difficult to form, and more difficult still to enunciate a just idea of this last number. Suffice it to say, that if from the time of the creation, dating it at 5841 years since, balls had been drawn incessantly from an urn, with such rapidity that one hundred millions had been taken every second, the number of drawings would only be represented by *nineteen figures.*

Supposing an urn of the magnitude of our globe filled with little balls of the diameter of ·000,000,078 of an inch only, (that is to say, infinitely smaller than grains of dust,) the number of these balls would only be represented by *forty-eight figures.* What is this then in comparison with all the chances which the drawing of two hundred balls present? Certainly, in the absence of theory, it would scarcely have been doubted that it were more easy to take, two hundred times in succession, a white ball from an urn containing equal numbers of white and black balls, than to lay hold of one single grain of white dust which might be contained in our globe, either on the surface or in the interior.

The consideration of compound events is very frequent where the probabilities of life are concerned. Thus we may be curious to know the probability that a man aged 40, and his wife aged 30, will both be living at the end of ten years. According to the Belgian Tables, which apply equally to France, the probability of living ten years, for men dwelling in towns, is, at the age of 40, 0·832 ; and for the wife, age 30, it is 0·862. The probability that they will both be living at the end of ten years will be the product of the two preceding fractions, which is 0·717. These numbers are only true in a general way, and it would be wrong to apply them particularly to such and such individuals: this I shall soon have occasion to explain.

If your Highness should desire to know the probability that after ten years the wife should survive the husband, you would observe that this event is composed of the two following,—that the husband will be dead, and the wife living. These two simple events have for their respective probabilities 0·168 and 0·862, the product of which is about 0·145.

Generally, four compound events are possible. I here give them, with their respective probabilities.

	PROBABILITY.
That the husband and wife will both be living . .	0·717
That the husband will be living, and the wife dead .	0·115
That the husband will be dead, and the wife living .	0·145
That the husband and wife be both dead . . .	0·023
	1·000

These four probabilities added together give unity, the symbol of certainty, which in fact should be the case. It is remarkable how feeble is the probability of the two consorts both dying in the ten years. This event has but two-hundredths of probability in its favour.

LETTER VII.

ON MATHEMATICAL EXPECTATION.—LOTTERIES.—ASSURANCE SOCIETIES.

WE are rarely permitted to witness the birth of a science, and take by surprise the first sign which reveals its existence: this, however, happened with the Theory of Probabilities. It was at a gaming-table that it was brought to light; and after having passed by frivolous amusements, it almost immediately attacked the most important problems which have occupied the attention of mankind. It undertook at the same time to direct observation, and to support it on more solid bases.

In its origin, the mission of the Theory of Probabilities was but to establish principles of equity between gamblers,—to determine the stakes, and to regulate the shares in case a player should quit the game before its termination. Now equity requires that two players should be placed in such a position that neither should have any advantage over the other. Thus, when they have equal chances of winning, they should stake the same sums. But *when the probabilities of winning are not the same, the players should stake sums proportionate to these probabilities.*

In a lottery, (in the Genoese Lottery, for example,) where there are 90 numbers, he who stakes on 5 of these numbers should stake five shillings; while the other player, who has in his favour the 85 remaining numbers, should stake eighty-five shillings: the stakes would be in the proportion of 5 to 85, or 1 to 17,—that is, in the ratio of the respective probabilities of winning.

This is easily understood. Suppose there are 90 equal chances and 90 players. Each player having but 1 chance will be in the same position as each of his neighbours, and will risk as much as each of them does,—one shilling, for example: his probability of winning is $\frac{1}{90}$. But 1 person may take the place of 5 of these players, by paying what they would have staked; and another person may, in like manner, take the place of the remaining 85, by also paying their stakes: the first should then pay five shillings, and the second eighty-five. These sums are exactly in the ratio of the chances, or the probabilities, the two players have of winning.

If the first player win, he would then receive for each shilling staked the sum of eighteen shillings,—that is to say, his own stakes plus those of the person with whom he plays. If he receives less than eighteen shillings the game is not equitable, and he is playing to his own detriment. This is the case with the Genoese Lottery: the player, instead of eighteen shillings to which he would be entitled, only receives fifteen,—the remaining three shillings form the profit of the proprietor of the game.

If, instead of playing on a simple drawing, he were to play on some fixed drawing, or on the drawing of a certain number at a certain defined place in the drawing, (the first, for example,) he would only have a probability of $\frac{1}{90}$ in his favour, and staking a shilling he should receive ninety in case of winning; but, instead of this sum, he only receives seventy,—the profit of the undertaker of the game thus being twenty shillings.

The "ambe," or the drawing of two specified numbers at one drawing, only gives 270 times the stake, while about 400 times the stake ought to be received; for with 90 numbers in the lottery 4,005 "ambes" can be made, by taking all the combinations 2 and 2 together, and with the 5 numbers which have been chosen 10 "ambes" can be formed. The probability of drawing an "ambe" is therefore $\frac{10}{4005}$.

The loss to the player, and consequently the benefit to the undertaker of the lottery, becomes still more considerable when the game is for the "terne," the "quaterne," or the "quine." The advantage to the undertaker of the game, which was but $\frac{3}{18}$ for the simple drawing, amounts for the "quine" to $\frac{42}{43}$ of the whole sum staked.

The name of *mathematical expectation* has been given to *the product of an expected sum multiplied by the probability of obtaining it.* Thus, when the stake is placed on the simple drawing with the view of receiving a sum of fifteen shillings, the mathematical expectation is calculated by multiplying fifteen shillings by the fraction $\frac{5}{90}$ or $\frac{1}{18}$, which expresses the probability of obtaining it: this gives $\frac{5}{6}$ of a shilling, or tenpence. Such is the sum then which we should equitably stake instead of a shilling.

In every kind of game, or of equitable bet, the mathematical expectation of the players should be equal. But your Highness has already been able to perceive how little, in the Genoese Lottery, the rules of equity are observed with respect to the player, and how dearly the owner of the game makes him pay for the pleasure of gaming. I have sometimes heard applied to lotteries what Buffon said of *pharaon,* " The banker is but an avowed rogue, and the punter a dupe whom it is agreed not to ridicule."

Games of chance yield perfectly to the application of the Theory of Probabilities; and each player may take account of his position, not only on entering the game, but at every stage of its progress : so that, if the players agreed to separate, they might know how they should share amongst themselves the sums at stake, taking into account their respective positions. In other respects these applications are more curious than useful.

But this is not the case when it relates to Assurance Societies. These societies differ essentially from gaming-houses in regard to morality. In both instances sums of money are placed at stake, but with very different views: the assurer is generally guided by motives of prudence and economy; the gambler, on the contrary, by improvidence and dissipation.

A father of a family, who does not possess a fortune, deducts from the modest produce of his labours some small sum, which he deposits annually, with the view of assuring an income to his widow should he die first. In order to regulate equitably the sum he ought to pay, recourse must be had to the consideration of compound probabilities.

Calculations of this description present no difficulties; but they are as various as the combinations to which Assurance Societies lend themselves are numerous. To take but one example: I will suppose that a married man, aged 40, wishes to leave to his wife, age 30, a sum of £1,000, in case she becomes a widow before the end of ten years. I have shown, at the close of my preceding letter that the probability of this compound event is 0·145. The £1,000. to be assured must then be multiplied by the probability of obtaining them,—that is by 0·145, which gives £145. the sum the husband ought actually to pay. This payment, though small, would be further diminished, if we take into consideration that the money deposited produces interest during the period of ten years. But, on the other hand, we must consider that the Assurance Society must make a profit for its shareholders, and for the expenses of management. This profit, in certain societies, is sometimes sufficiently exorbitant to resemble that made by the undertakers of lotteries; and although the opposite natures of these two institutions permit us to suppose that different sentiments actuate those who come to risk their money in them, they may still be each guilty of the same vices, by an immoderate excess of gain which might justly be called usury.

I should perhaps here speak of a difficulty experienced in the countries where pension-funds have been established for the widows and orphans of public officers. Generally, not only the married, but the unmarried men also, have been made to contribute to these funds; and it is perfectly right, although the latter have often considered as unjust the contribution imposed upon them.

In what manner is a public functionary affected? By a pension eventually payable to his widow. But the unmarried man will say, "I have no wife, and am not therefore in a position to leave a widow." The married man may reply, "Although I have a wife, it is as possible that I shall survive her, and the fund not have to pay the annuity, as it is that you, who are not now married, will take to yourself a wife, who will survive you, and to whom the pension will become due. Each of us then ought to pay for the probability of leaving a widow."

Now, for the married officer, we must consider the probability of his dying before his wife,—a probability which will regulate the sum he should pay in order eventually to leave a pension. For the bachelor, the payment of a pension to his widow depends on a compound event: we must consider the probability of his marrying, and the probability that, being married, he will die before his wife. These two probabilities can only be estimated in a general way,

and from the results of many years of observation in the administration of public affairs.   Their product afterwards regulates the
sum to be paid annually, in order that the bachelor, in case of his
marrying, may participate fully in the advantages of the fund.
Only after his marriage the annual contributions will become
larger, because the chances of leaving a widow are greater.

The annual contribution of the bachelor might in fact be dis-
pensed with, provided he be made to pay a sufficient sum at the
time of his marriage,—the date from which he takes a share in the
benefits of a fund already established, and provided by the savings
of his colleagues.   But such a sum, paid at the time of marriage,
would be a heavy charge, the weight of which would be less felt
when distributed equally over the whole term of his bachelorhood.

It is vain to pretend that all the bachelors will not marry; it
may as reasonably be objected that all the married officers will
not die before their wives : the one as well as the other makes a
present sacrifice eventually to obtain a benefit, — sacrifices which
ought to be proportional to their mathematical expectations.

It is both consolatory and moral that men should unite and
assist one another, to combat the scourges which menace them
incessantly,—and that, by means of small sacrifices in the days of
prosperity, they should husband for themselves powerful resources
against times of misfortune.   It is the accomplishment of one of the
first duties of Christian charity.   We should think kindly of the
science which tends to regulate this duty according to the principles of justice, and which studies the means of making it produce
the most useful results with the least possible sacrifice.

I will beg leave to call your Highness' attention to another
advantage which is presented by Assurance Societies and Savings'
Banks : this advantage concerns the government much more than
individuals.   These latter, in assuring, present the state with a sort
of guarantee that they will respect public order : they will not, in
fact, compromise the prospects of their families, by exposing without reason the produce of their economy to the chances of political
revolutions.   I am often astonished that governments should not
take a more direct part in institutions which may so advantageously
develop the spirit of order, and the morality of a nation.

LETTER VIII.

THE wise and prudent man avoids games and bets, even when he is sure of seeing all the rules of the strictest equity observed, and particularly if considerable sums are hazarded. This is because by the side of the mathematical question is presented one of a superior order. Should we not be right, in fact, in showing to a friend who is exposing half his fortune what may be the consequences of his imprudence? Ought we not to make him feel that the privations which he would have to impose upon himself, in case of loss, could in no way be compensated by the advantages he would gain were he to win? It is not sufficient for mathematical principles to be strictly observed,—moral ones should be equally regarded. If a man were to present me with a loaded pistol, inviting me to toss up for the purpose of deciding which of us should fire on the other, I should certainly take him for a madman, and would not accept his proposition, although he might pretend that the mathematical probability of being killed were exactly the same for each of us.

When sums of money are put at risk, the evil results of loss are doubtless not so great: it however may in certain cases compromise our existence, our honour. In general, that only should be exposed to risk which may be lost without inconvenience, and without doing injury to the fortune of the player; in other words, we should only expose sums small in comparison to our possessions.

"The miser is like the mathematician," said Buffon, in his *Essay on Moral Arithmetic,*—"both estimate money by its numerical quantity. The man of sense considers neither its mass nor its number: he only sees in it the advantages which he may obtain from it; he reasons better than the miser, and perceives better than the mathematician. The crown which the poor man has set aside to meet a payment imposed upon him by necessity, and the last crown which fills the bags of a capitalist, have to the miser and the mathematician but one and the same value: the latter will count them by equal units,—the other will appropriate each to himself with an equal pleasure; while the man of sense will consider the poor man's crown a pound, and the crown of the capitalist but a mite." It is in this sense, in reality, that it becomes right to estimate the importance of a sum by a comparison with one's entire possessions; and *this importance, which is called the moral value, is valued by dividing the sum by the amount of property possessed by the person who exposes it.* Thus £100. has to him who possesses but £200. the same moral importance as £500,000. has to him who possesses £1,000,000.

Adopting this method of calculation, it is easily perceived that the most equitably regulated game can but be unfavourable to one or other player. To make it more apparent, I will suppose that a person who has but a £1,000. exposes £500. at a game of toss. The moral value of this sum will be $\frac{1}{2}$, while the moral value of the £500. which he obtains in case of winning will be represented by $\frac{1}{3}$ only. Thus, in relation to the £1,000. that this person possesses, the importance of the sum set at risk is represented by £500, and that of the sum gained by £333. 6s. 8d.: there is then a difference of £166. 13s. 4d. between the stake and the profit. This is the moral detriment to which the player is exposed.

This detriment may in truth become insignificant, when the money risked forms but a small portion of the player's property; the difference, also, between the moral values of the sum hazarded and the sum played for may then be considered as insensible.

A difficulty may arise in regard to the appreciation of the importance of a sum to an individual who actually possesses nothing. But it must be remarked, that for him who lives by the work of his hands, this work itself represents a capital which constitutes his fortune. He who begs to maintain existence finds resources in his condition, which he would not exchange for a like sum to be paid annually in the form of a fixed revenue. There are but few, save the unfortunates who are dying of hunger, who can be considered as possessing absolutely nothing.

The same considerations should lead us never to expose our whole fortunes on the chances of a single occurrence. If I had a considerable sum to transport to America, I should not place it all in one vessel, as in case of shipwreck I should be ruined. If, however, I distribute the sum among many vessels, each having the same chance of being destroyed, I run the risk of partial losses, but I have little fear of the realization of the compound event, on which would depend a total loss.

Prudence should, in like manner, guard us against placing our whole fortunes in the hands of one person, whatever may be the guarantees he may present.

I feel persuaded that your Highness, for like motives, would see with regret all the inhabitants of the several provinces occupying themselves exclusively with the same species of cultivation. Were this mode of culture to fail, the effect would be very perceptible; whilst by varying the different modes of culture, the chances of the failure, at any one time, of all the different crops of a year are extremely small: the evil is thus but partially felt, since the nourishment of the poor has become more varied, and has comprehended a greater variety of produce, and a greater certainty of its being gathered in,—the scarcities which so often afflicted our forefathers have become almost impossible. This is a benefit, at the same time that it is a consequence of the progress of civilization.

Buffon, whose writings have contributed much to throw light on the subject which now occupies our attention, gives the following example to show that in games, even those which are the most fair, viewed mathematically, the positions of the players are not equally favourable.

"If two men were to determine to play for their whole property, what would be the effect of this agreement? The one would only double his fortune, and the other reduce his to nought. What proportion is there between the loss and the gain? The same that there is between all and nothing: the gain of the one is but a moderate sum,—the loss of the other is numerically infinite, and morally so great that the labour of his whole life may not perhaps suffice to restore his property.

D

"The loss then is infinitely greater than the gain, when all our wealth is staked: it is a sixth greater when we risk half our property: it is a twentieth greater when we risk a quarter of our property. In a word, however small the portion of our fortune that we hazard at play, there is always more loss than gain: the compact also of the game is an erroneous contract, and tends to the ruin of the two contractors.—A new but useful truth; and how much do I desire it to be known by all those who, from cupidity or indolence, pass their lives in tempting fortune!

"The question has often been asked, why we are more sensible of loss than gain? A fully satisfactory answer could not be given, so lông as we were in doubt of the truth I have just presented. Now, the answer is easy. We are more sensible of loss than of gain, because, in fact, supposing them numerically equal, the loss is nevertheless always and necessarily greater than the gain: the feeling is in general but an implicit reasoning, less clear, but often more subtle, and always more certain, than the direct product of reason. We are well aware that gain does not give us so much pleasure as loss does pain: this feeling is the implicit result of the reasoning I have just presented."

These considerations would be well fitted to moderate the cupidity of gamers, were they to consult reason in the slightest degree before tempting fortune.

There are parts more disastrous, which are sometimes undertaken with the most culpable temerity,—they are those that are played in battle-fields. Sad conditions are those which exact as the stake the prosperity of nations and the blood of men,—where the winner is he who best succeeds in destroying what he covets, without being able to foresee the extent of the sacrifices he must impose upon himself to obtain it!

LETTER IX.

I HAVE now the honour to write to your Highness on an important question,—that of judging of the degree of confidence to be placed in a result given by the Theory of Probabilities.

We should be strangely mistaken, if we believed that experience would always justify the anticipations of calculation. This accordance is in general but accidental; but we can make the difference as small as we wish, regard being had to the nature of the researches with which we are occupied. Before we examine this important point, with all the care and detail that it merits, it may not be amiss to put ourselves on our guard against the abuses which we might be tempted to make of the theory.

When but a single trial is made, we may never have an agreement between the result of calculation and that of experience. Before the event, there are but probabilities for or against its occurrence; and when the event is produced, probability is replaced by certainty. Thus, before taking a card from a pack, I have $\frac{12}{52}$ as the probability of drawing a picture-card, and I have against me the probability $\frac{40}{52}$: however, when the card is drawn my position is entirely changed,—my doubts are replaced by certainty; and if I have betted upon the event, I know whether I have won or lost, whatever may have been the probability in my favour.

When a number of trials is made, an agreement may be established between the results of calculation and of experience; but this agreement is not necessary. James Bernoulli has shown that, *by multiplying sufficiently the number of trials, we may attain to a probability as near to certainty as we may desire,—that the difference between the results of calculation and of experience may be contracted within as narrow limits as may be desired.*

This is well known to those who establish lotteries or gaming-houses. The great number of players who go there to risk their money, notwithstanding the apparent fluctuations of chance, leave to the undertaker of the game a profit which they can calculate in advance, and which is as fixed as the revenues of the Treasury. Thus *Les Recherches Statistiques sur Paris* informs us, that from 1816 to 1820 inclusive, the Paris Lottery put in circulation £1,000,000. annually, of which sum the Treasury received rather more than one-fourth.

The Belgian Government, in instituting pension-funds for the widows and orphans of public officers, has unfortunately lost sight of the important principle, that calculation can only agree with experience in proportion as it operates on large numbers. By forming separate funds for each branch of public service, and by imposing on itself the condition of never coming to their aid, it has multiplied the chances of fluctuation to which those funds are necessarily subject. It becomes but little probable that the hopes of a father will be realized, with respect to a pension to his widow and orphans, when an association does not number a hundred members, as would be the case were the numbers greater. The natural consequences of such a state of things is, that certain funds will prosper while others suffer. This inequality will necessarily be a great evil, as it will differently affect men who ought all to be considered as members of the same family.

It is a very simple and very useful principle in practice, that *the precision of the results increases as the square root of the numbers of observations.* Thus, *cæteris paribus*, the degrees of precision are as the numbers 1, 2, 3, 4, &c., when the observations are as the numbers 1, 4, 9, 16, &c.

I have been induced to test this principle by experiment. I threw into an urn 20 black balls, and an equal number of white balls; so that the probability was the same (viz. $\frac{1}{2}$) of drawing a black or a white ball. It appeared then, that after a certain number of drawings, the number of white balls drawn should be equal to the number of black balls: this, however, was not the case, as may be seen by the following table, which indicates the results successively obtained after 4, 16, 64, &c. drawings. I should mention that, after each drawing, the ball drawn was replaced in the urn, that all the conditions might remain the same.

Number of Balls drawn.	Degree of Precision.	Number of White Balls.	Number of Black Balls.	Ratio of the preceding Numbers.
4	2	1	3	0·33
16	4	8	8	1·00
64	8	28	36	0·78
256	16	125	131	0·95
1024	32	528	496	1·06
4096	64	2066	2030	1·02

The first column shows the number of balls drawn,—the second the square roots of the same numbers: according to the principle enunciated above, these roots express the relative degrees of the precision of the results. In the two next columns are shown the numbers of white and black balls drawn from the urn: these numbers should be equal, if theory and experience rigidly agreed. Now the last column shows the ratio of the number of white balls drawn to the number of black. This ratio should be 1; but this result was only obtained once, and then after but 16 drawings. This accordance was but accidental; whilst we remark that, in the oscillations of the numbers, a very evident tendency to approach to unity is exhibited as the number of drawings is multiplied.

If your Highness is astonished that I insist upon so common an example, I might in justification cite a high authority,—that of the illustrious Buffon, who, in his *Essay on Moral Arithmetic*, cites similar trials which he caused to be made by a child.

SECOND PART.

ON MEANS AND LIMITS.

LETTER X.

ON MEANS AND LIMITS IN GENERAL.

WHEN we stand in the presence of Nature, and seek to interrogate her, we are at once struck with the infinite variety which we observe of the least phenomena. Whatever may be the limits within which we concentrate our attention, we find a diversity as astonishing as it is embarrassing. The most simple appreciations leave a vagueness incompatible with the precision which science requires. One single object, measured or weighed several times in succession, notwithstanding every precaution that may be taken, nearly always presents dissimilar results. Our ideas, however, seek to fix themselves, and to settle on a precise number,—on some mean which will show the results of the observations made, as free as possible from accidental error.

The consideration of means is so familiar to us that, wherever we meet objects, which fix our attention and are subject to variation, we employ it in some measure without our own knowledge. Thus we have no difficulty in attributing to the sun an apparent determinate size, although that luminary is not seen exactly under the same angle two successive days, or we may say two successive instants. In the same manner we state the temperature of summer in given places, although this temperature varies every instant.

The Theory of *Means* serves as a basis to all sciences of observation. It is so simple and so natural that we cannot perhaps appreciate the immense step it has assisted the human mind to take. We are ignorant to whom it is due: it is thus that many great discoveries have been established, without their discoverers being known. All that we learn from the history of science in this respect is, that one people made use of the discovery before another. This is the case with numeration, with writing, and even with printing.

Let us here remark, that while prepossessed with the idea of the mean of variable quantities, the *limits*, between which these variations operate, have perhaps been too much lost sight of. In all things to which *plus* or *minus* may be applied, there are necessarily three things to consider,—one mean, and two extremes.

Without having recourse to science, we gain from habit a vague appreciation of the mean and of the limits of variation which belong to each variable element, presented either in nature or in our social condition. We are guided in our reasonings accordingly. But it is convenient for the progress of enlightenment to substitute precise ideas for vague notions.

It would be curious to inquire at what epoch a rational use was first made of *means*. There assuredly existed an obscure idea of them in ancient times, for the idea, as I have already remarked, is inherent in our nature, and serves as the basis of nearly all our opinions; but it was not explicitly produced until a late period, and it would be difficult to fix the epoch of its introduction in modern times, or to say from what date it became an established principle that *the mean of a series of observations is obtained by dividing the sum of the values observed by the number of observations.* The consideration of limits, inasmuch as they complete the idea of the mean, could only be established by the application of the calculus of probabilities to the study of natural phenomena. The establishment and the development of the Theory of Means would form one of the most interesting chapters in the history of the human mind.

Archimedes, that genius remarkable in so many respects, seems to have best appreciated in ancient days the importance of means: he made an admirable use of them in his researches on the centre of gravity, of which he was the discoverer. He substituted the consideration of one point for that of a great number; and this ingenious idea, which has since been so fruitful, alone makes him worthy of the gratitude of mankind.

When Meton, four or five centuries before the Christian era, presented to the Greeks assembled to celebrate the Olympic Games his famous period, which was received with so much enthusiasm, and engraved in letters of gold, he could only discover the length of a lunation, or of a tropical year, by a calculation analogous to that of means. The mind, like the body, walked with a firm step long before the laws of equilibrium were known.

Aristotle, one of the greatest of ancient observers, also perceived the properties of means: he applied them to the moral sciences. Virtues, according to him, consist in an exact state of equilibrium; and all our qualities, in their greatest deviations from the mean, produce vices only.

This doctrine passed from the school of the philosophers to that of the poets. Horace, among the Romans, was one of the most elegant and happy of its interpreters.

But how great is the distance between these first perceptions and the learned theories which we possess in the present day! However, there yet remains a great distance to be traversed before the brilliant labours of modern analysis produce their full fruits. The greater number of observers, even the best, know but very vaguely—I will not say the analytical Theory of Probabilities, but the part of this theory which relates to means. I hope, then, that the time we shall employ on it will not furnish matter for regret.

LETTER XI.

ON MEANS PROPERLY SO CALLED, AND ON ARITHMETICAL MEANS.

IN the spot where your Highness dwells, the return of Spring should doubly charm. The prospect of a great river is a spectacle as imposing as it is varied,—that of the Rhine in particular pleases the imagination, which there finds, in some measure, the personifications of Ancient Germany in its days of glory, and in its days of misfortune. The cloudy tops of the seven mountains, which are visible on the horizon, form a majestic boundary to the imposing scene, and recall the enchantments of the middle ages.

I should wish to reside in Bonn on the return of Spring, and to pass my days on the borders of the Rhine. I would renounce all study, my mind only occupied with the animating scene which that fine river presents. I see here boats crossing in all directions, and travellers of all nations coming to pay their homage to one of the fairest and richest valleys in the world. These waves, which flow so rapidly before me, whence come they? A few days since they descended perhaps from the summit of the Alps; and they now go to lose themselves in the bosom of the ocean. To judge by the rapidity and breadth of the river, the quantity of water which passes must be enormous: it would seem that the Diluvian rains only could supply so vast a channel.

But while I speak of fleeing from study, and seeking repose, I find myself led back (spite of myself) to my former habits. I already calculate the velocity of the water, and ask myself, "Is this velocity always the same?" and supposing it variable, "What is its average rate?" If I carry my thoughts further, and consider the total quantity of water collected in the channel through which the Rhine runs,—if I seek to learn how much water should fall through a square mile of surface to feed this river, I find myself still met by averages. Throughout, I am forced to have recourse to their use. It is important to acquire correct ideas of their theory.

In taking a mean, we may have in view two very different things : we may seek to determine a number which actually exists, or to calculate a number which gives the nearest possible idea of many different quantities, expressing things of the same nature, but various in magnitude. To explain,—

In measuring the height of a building twenty times in succession, I may not perhaps twice find the same identical value. However, it may be conceived that the building has a determinate height; and if I have not exactly estimated it, in any one of the operations I have made to discover it, it is because these operations are liable to some uncertainty. I content myself, then, by taking the average of all my results as the true height sought. The limits, greater or smaller, depend on my skill or want of skill, and on the exactness of the instruments which I have used.

I may employ the calculation of the mean in another sense. I wish to give an idea of the height of the houses in a certain street. The height of each of them must be taken, and the sum of the observed heights must be divided by the number of houses. The mean will not represent the height of any particular one; but it will assist in showing their height in general,—and the limits, greater or smaller, will depend on the diversity of houses.

There is between these two examples a very remarkable difference, which perhaps may not have been seen at first glance, but which is nevertheless of great importance. In the first, the mean represents a thing really in existence; in the second, it gives, in the form of an abstract number, a general idea of many things essentially different, although homogeneous.

In another view (and this point is important) the numbers which have contributed to form the mean in the two examples present themselves in very different manners. In the second example, they are bound to one another by no law of continuity; while in the first, as we shall soon have occasion to see, the determinations of the heights, although faulty, range themselves on each side of the mean with so great a regularity that their values might be predicted, if the limits within which they are comprised were given.

This distinction is so important that it should not be lost sight of. I will put it into other words, to illustrate it better. I will reserve the name of *mean* for the first case, and will adopt that of *arithmetical mean* for the second, in order to show that reference is here had to a simple operation of calculating quantities which have no essential relations. These relations are not always perceptible: they are frequently recognised where they were not expected to have been found. The arithmetical mean then becomes a true mean.

Sometimes the arithmetical mean is deduced from most different elements, without our being able thence to conclude that the general idea which it should represent may be useless or unimportant. I will quote as an example the mean duration of life. We know that the statist, when he wishes to calculate it for a given country, supposes that all the individuals born at the same time in that country put together the years, months, and days they have to live, and make an equal division amongst themselves, so that one shall not live longer than another. Thus, of 100,000 individuals, 9,600 live only one month, 2,460 live two months, 1,760 live three months, and so on. The total of the duration of life of these individuals is taken, and divided by 100,000,—the result is the average duration of life. It is about thirty-two years for Belgium and France: in England it reaches thirty-three years.

It has been remarked that, according to the progress of civilization and science, the average duration of life has become greater among some nations; and it is a subject for congratulation, that it is most frequently the case that the increased duration of life is shared by all individuals.

It is, however, well to remark that the consideration of the mean duration of life often leads to error: this may be proved from what has just been said. For example,—from the mean duration of life being the same in France and in Belgium, it must not be concluded that these two countries are in the same circumstances as regards mortality.

If ten years of age were removed from an old man, the father of a family, to be added to the life of one of his infants, who had died immediately after its birth, the average duration of life would in no wise be changed; but this transfer of ten years would have an immense influence on the fortune of the family, which, having lost its prop and its support, would at the same time have one more to nourish.

The number representing the average duration of life gives but a general idea of the mortality, and can only be employed with circumspection. It would be difficult to cite any example of arithmetical mean in which more dissimilar elements are employed. In the calculation of the average duration of life, the same value is given to a year of existence of a child as of a man in the prime of his life or in his old age.

LETTER XII.

THE sciences of observation—political sciences in particular—
frequently require the use of means. The statist may wish to
know the price of labour, and of the principal objects of consumption; but these prices may vary from day to day, and sometimes
in adjacent places. It is important, then, to know what is the
mean, and what the extreme prices.

If the value of wheat is under consideration, we should not be
satisfied with considering such and such a locality, or such and
such a season of the year, or even such and such a year in particular, but should seek a general arithmetical mean, which would
enable us to embrace a certain number of years, and we should
at the same time inquire within what limits this mean was comprised. Thus, going back to the origin of the Kingdom of the
Netherlands, we find for the two decennial periods, and the sexen-
nial period which followed,—

Period.	Mean Price.	Maximum Price.	Minimum Price.
	s. d.	£. s. d.	s. d.
1817 to 1826	14 2	1 8 4 in 1817	8 10 in 1824
1827 „ 1836	14 6	0 18 10 „ 1829	10 6 „ 1834
1837 „ 1842	16 4	0 18 2 „ 1839	13 0 „ 1837

A glance at this table shows us many important facts,—

1st. The average price of wheat increased progressively from 1817 to 1842.

2nd. The limits between which the prices varied have successively contracted.

For the first decennial period the maximum and minimum prices differed 19s. 6d.; for the second period 8s. 4d.; and only 5s. 2d. for the third period, which only extended over six years.

But the two facts I mention are a natural consequence of our social state. The value of money progressively decreases when referred to useful and indispensable articles, such as wheat. This fact is well known to economists, who have announced the causes of it: it is so manifest that the observations of a few years are sufficient to prove it.

The second fact is not less important, nor less curious, although it has been less noticed. It is a general consequence of the progress of civilization, that all social elements subject to variation oscillate between limits, so much the more narrow as our knowledge of them is more advanced: and by " our knowledge," I mean not merely actual acquaintance, but also the wisdom of our institutions,—political calm,—and all that can preserve the citizens of the state from the scourges which threaten their persons and their goods. I have endeavoured to show elsewhere* all the advantages which result from narrowing the limits within which the different social elements may vary.

To take but one example. The preceding table shows us that the greatest variation in the price of wheat took place in 1817: the price rose to £1. 8s. 4d.,—that is, to double the mean price. This calamitous year has left deep and deplorable traces in the mortuary tables, and in all elements which relate to the well being of mankind.

A great fall in the price of grain would not produce less grievous results; only the evil would be felt in another quarter, and would affect the producer instead of the consumer. If there were suddenly to be introduced into Belgium, from abroad, sufficient grain to reduce the ordinary price one-half, it would be a real

* *Physique Sociale*, 2 vols. Paris: Bachelier, 1835.

calamity to the agriculturist, and would not be long in showing its effects, especially in the mortuary lists. If this fact has been less observed than the other, it is because the differences of the prices of grains do not vary to an equal extent on each side of the mean, the variations of increase being generally greater than the variations of decrease.

The statesman, in the habit of treating general questions of public utility, is forced to recur to the use of *means* more frequently than other men. This habit suggests to him, in order to arrive promptly at his conclusion, expedients more or less ingenious, more or less easy in practice. In the absence of exact documents, scientifically collected, he must often content himself with statements which only approximate to the truth.

A law affecting the interests of industry and agriculture could not be satisfactorily prepared, without connecting with it questions of statistics: nevertheless, even the most civilized countries scarcely know their resources, or their wants.

LETTER XIII.

MEANS also are much employed in the physical sciences,—in Meteorology particularly. Thus, to form an idea of the temperature of any one day, the arithmetical mean between the two extreme temperatures of such day would be taken. If it were found that during the night the thermometer had fallen to 42° Fahrenheit, and that it had afterwards risen to 60° towards the middle of the day, we should say that the mean temperature had been 51°.

This mean value may give an idea of the temperature, although it is insufficient in many respects, for it might have been produced in different manners. The temperature might have been uniformly 51° throughout the twenty-four hours; or it might have oscillated between the limits of 42° and 60°, or even between wider limits.

The consideration of the limits is not unimportant, since it gives us the measure of the variations to which the temperature has been subject; and these variations, when they are rather extensive, might be very injurious to health. It is useful to consult them, when it is desired to form an idea of the salubrity of a climate. It has been noticed that they are in general less sensible at the seaside than in inland places. They are not, nevertheless, the same in a given place during the different seasons. A short discussion of such a subject will be well suited to give somewhat correct ideas of the question under consideration; for the mode of procedure in this kind of research is applicable to an infinite variety of circumstances.

Let us suppose it to relate to the study of the temperature of Brussels. We will first call to mind that the mean temperature of a day is the arithmetical mean of the several temperatures exhibited by observations made every instant during the day. But the learned, even the most persevering, would decline to make use of it, were it only possible to determine it at this expense. Happily, it has been discovered that it can be arrived at by processes different to that which I have just pointed out, which would require the presence of the observer at his instruments during twenty-four successive hours, without allowing him an instant's relaxation. One of these processes consists in taking the arithmetical mean of the two extreme temperatures during the day; and the number thus obtained differs little from that which would have been obtained by continued observations. Moreover, it is not even necessary to be near the thermometer, to watch the value and the instant of the *maximum* or of the *minimum*. We possess instruments which register of themselves the two extreme temperatures; and the observer has only to transcribe the results, and to adjust the instrument so that it may furnish further observations on the morrow. It is by these simple means then that we obtain (very nearly at least) three important things,—the mean and the two extreme temperatures of the day.

Let us first pause at the former of these data. In taking it during several successive days, we soon perceive that the mean temperature is very variable. In 1842, for example, in the month of July, the mean daily temperature had different values between the extreme limits of 77° and 53°. The arithmetical mean of the thirty-one different temperatures during the thirty-one days of July gave 63°: this quantity represents the *mean temperature of the month*. If curiosity were to lead you to ascertain, by reference to the publications of the Observatory at Brussels, what temperature obtained during the preceding years, you would find for the decennial period from 1833 to 1842 as follows. In addition to the mean daily temperature, I have given the temperature of the hottest and of the coldest day of the month of July in each year.

E

TEMPERATURE OF JULY AT BRUSSELS.			
Year.	Mean Temperature.	Hottest Day.	Coldest Day.
1833	65	70	62
1834	70	78	58
1835	66	74	53
1836	65	72	48
1837	63	72	54
1838	65	79	54
1839	65	71	52
1840	61	67	55
1841	59	68	56
1842	63	73	57
MEAN . .	64	72	55

The mean temperature of a July day is then 64°, consulting the means of the several days of July during the decennial period from 1833 to 1842,—that is 310 means, the highest of which was 79° in 1838, and the lowest 52° for 1839. It is between these two extremes that the daily temperature has been constantly included. But it is extremely improbable that these two limits would occur in one year. The most probable limits are those deduced from the totals of the particular values of each year, that is 72° and 55°.

In regarding 64° as the arithmetical mean of 310 observations of daily temperatures, made during the month of July in each of ten years, I have not supposed any necessary relation between these 310 numbers. It might, however, happen that such did exist without my knowledge, and that all these numbers were not presented without a certain order. It may be interesting to investigate this.

The most simple way to verify such a relation, if it exist, is to arrange impartially all the temperatures observed, according to the order of magnitude, classing them for instance in degrees. This is done in the following table :—

Degrees of Temperature.		Number of Days each Year.										The 10 Years.
Centigrade.	Fahrenheit.	1833	34	35	36	37	38	39	40	41	42	
11 to 12	51·8 to 53·6	...	...	...	...	...	...	1	...	...	...	1
12 „ 13	53·6 „ 55·4	...	...	...	1	1	5	1	1	...	...	9
13 „ 14	55·4 „ 57·2	...	...	1	1	...	...	3	2	13	1	21
14 „ 15	57·2 „ 59·0	...	1	...	3	2	1	...	4	7	6	24
15 „ 16	59·0 „ 60·8	2	...	1	3	2	5	1	6	5	8	33
16 „ 17	60·8 „ 62·6	3	1	4	3	8	2	2	6	1	2	32
17 „ 18	62·6 „ 64·4	7	...	3	5	10	4	6	6	2	6	49
18 „ 19	64·4 „ 66·2	7	3	4	3	5	3	3	4	1	2	35
19 „ 20	66·2 „ 68·0	7	5	4	4	...	1	5	2	2	1	31
20 „ 21	68·0 „ 69·8	3	3	8	2	2	2	3	...	...	1	24
21 „ 22	69·8 „ 71·6	2	7	4	1	...	3	3	...	...	1	21
22 „ 23	71·6 „ 73·4	...	4	1	4	1	3	3	...	...	1	17
23 „ 24	73·4 „ 75·2	...	3	1	1	...	...	...	...	...	2	7
24 „ 25	75·2 „ 77·0	...	1	...	...	...	1	...	...	...	...	2
25 „ 26	77·0 „ 78·8	...	3	...	...	...	...	...	...	...	...	3
26 „ 27	78·8 „ 80·6	...	...	...	...	...	1	...	...	...	...	1

The temperature most often repeated is that between 17° and 18° of the centigrade scale, or 62·6° and 64·4° Fahrenheit, which is found to have occurred on forty-nine days; and it is to be remarked that this is nearly the mean temperature given in the total of the observations, *i. e.* 64° Fahrenheit.

The next temperatures which occur most frequently are those which differ the least from this latter: we may even see that they rank almost symmetrically, in order of magnitude, on the two sides of 49. Physicians render these abstract results sensible to the eye by means of geometrical construction, or by a curve, which shows at a glance the progress of a series of numbers.

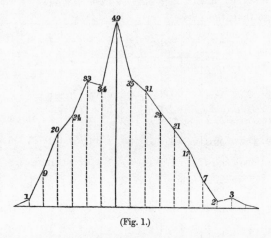

(Fig. 1.)

The line traced in the figure furnishes an example. With some few exceptions, we see that this line rises in a very symmetrical manner, until it attains a *maximum* value, and then decreases. Its form is in general very irregular, when it refers to an arithmetical mean of numbers which have no necessary relation one with the other. But when the mean is not a purely abstract numerical result, and there exists, in relation to the season and the locality, a general temperature (which may, however, be more or less affected by fortuitous circumstances), the curve exhibits it by its regularity, provided the number of observations be sufficient. This appreciation is very delicate. I propose to refer especially to it in my future letters, but at present I confine myself to mentioning it.

To return to our first example. I have said that there are three important things to consider in the discussion of the climate of a country: these are the mean and the two extreme temperatures of each day. This variation is not the same for every day,—it may be influenced by many atmospheric circumstances; but when we combine the observations of a whole month, and take the arithmetical mean of the variations observed in the thirty-one successive days, we obtain a quantity more or less free from the effects produced by fortuitous causes, which quantity represents the normal effect of the season with respect to the climate. This variation, determined daily during a certain time, is much less subject to change than the mean temperature. What has been observed one year is obtained very sensibly in the following years. To show this, I will construct a table, analogous to that given before, for the daily temperatures.

VARIATION OF THE TEMPERATURE DURING 24 HOURS IN THE MONTH OF JULY.			
Year.	Mean Variation.	Maximum.	Minimum.
1833	18·7	32·0	12·8
1834	18·2	27·0	12·1
1835	22·7	29·3	16·3
1836	18·4	27·5	12·1
1837	18·0	26·8	11·3
1838	17·1	28·4	8·8
1839	16·9	26·8	8·6
1840	14·9	22·3	7·2
1841	14·2	23·7	7·4
1842	18·4	25·2	6·1
MEAN . .	17·8	27·1	10·2

The variations of temperature during a July day are, then, nearly 18° Fahrenheit: this value has changed but little from year to year. However, in considering the days individually, I find that it once rose in 1833 to 32°, and once in 1842 it was but 6·1°. The interval between these two extreme variations is 25·9°. The mean temperature of a day during the month of July has not passed the two limits 79° and 52°: the interval passed through then is 27°, which differs but little from 25·9°.

As to the manner in which the 310 variations of temperature, observed in July during ten consecutive years, are distributed, we may judge by the following table, which includes also the variations of temperature for the month of January during the same period of time.

VARIATIONS OF TEMPERATURE IN 24 HOURS.			
DEGREES.		JULY.	JANUARY.
Centigrade.	Fahrenheit.		
From 1 to 2	1·8 to 3·6	...	8
2 ″ 3	3·6 ″ 5·4	...	31
3 ″ 4	5·4 ″ 7·2	1	61
4 ″ 5	7·2 ″ 9·0	7	68
5 ″ 6	9·0 ″ 10·8	10	50
6 ″ 7	10·8 ″ 12·6	24	32
7 ″ 8	12·6 ″ 14·4	34	22
8 ″ 9	14·4 ″ 16·2	41	20
9 ″ 10	16·2 ″ 18·0	42	8
10 ″ 11	18·0 ″ 19·8	41	4
11 ″ 12	19·8 ″ 21·6	38	3
12 ″ 13	21·6 ″ 23·4	26	1
13 ″ 14	23·4 ″ 25·2	19	1
14 ″ 15	25·2 ″ 27·0	16	
15 ″ 16	27·0 ″ 28·8	6	
16 ″ 17	28·8 ″ 30·6	3	
17 ″ 18	30·6 ″ 32·4	1	

The mean variation for July has been little less than 18° Fahrenheit, as we have seen. It is again the case with this variation that the greatest number of observations correspond. The other numbers rank nearly symmetrically on the two sides of this mean.

For the month of January the mean variation of the temperature was 9·3°. We may now remark that the numbers do not group themselves symmetrically on the two sides of the mean. We may perhaps see the reason hereafter.

It will have been remarked how much greater the daily variation of temperature is in July than in January. This is not owing to accidental, but to constant causes. In making the calculation for each month individually, we see in fact the daily variation progressively decreases in proportion to the distance from the hottest month; and it is at its *minimum* in January, the epoch of the greatest cold in winter.

The following table will furnish further information in this respect :—

MONTHS.	Mean Temperature.	Daily Variation.	Monthly Variation.	Greatest Variation in 10 Years.
January .	35·24	9·18	26·64	37·62
February .	39·38	10·26	23·40	41·52
March . .	42·80	12·06	25·56	42·12
April . .	47·30	14·94	28·08	44·10
May . . .	52·20	18·18	31·68	46·80
June . . .	63·32	18·36	32·22	45·36
July . . .	64·40	17·82	30·06	41·04
August . .	64·40	17·64	30·96	45·54
September .	59·36	14·94	27·72	41·76
October . .	41·80	12·24	24·84	36·72
November .	43·70	9·90	23·76	40·86
December .	39·38	8·82	22·68	43·02

The first column shows the mean temperature of each month, and the second the variation this temperature undergoes in the space of four and twenty hours. The third column indicates the average variation of the temperature during a month : it is obtained by comparing the two extreme temperatures of the month, which gives ten values for the ten years of observation; and in the table is given the arithmetical mean of these ten numbers. Lastly, the fourth column gives for each month the difference between the two extreme temperatures observed during the space of ten years.

The daily variation increases with the increase of temperature, but it would be difficult to say after what law. It is, all other things being equal, a little greater during the first six months than during the second.

It would, moreover, appear that this variation depends on the length of the days: it attains its *maximum* in June, and its *minimum* in December. During this latter month it is nearly half what it is in the summer.

It can easily be seen that the temperature in decreasing and passing below zero would induce variations of temperature smaller and smaller, and which would probably become nought (if we admitted the law of continuity) long before reaching a temperature equal to that of the planetary spaces, which would nearly correspond to 74° below zero. The theory does not admit as necessary the absence of daily variation, but in the case where the earth may have a temperature equal to that of the planetary spaces: it would then be in a steady equilibrium of temperature, and would receive from the celestial bodies exactly as much heat as it radiated to them.

We might be curious to know also what would be the other limit of the daily variation of temperature,—that is to say, where it would attain its greatest value. The *maximum* would occur, were we to suppose the sun to be passing our zenith, and the earth to be without an atmosphere. The rays would reach us vertically at noon,—they would not be partly extinguished by the interposition of the air; and in the night radiation would be at its *maximum*.

But the temperature, which in summer rises at intervals to 90°, might, if the sun were in the zenith, rise to 104° or even 125°; and as the air (which in such a case absorbs at least one-fourth) would no longer exist, this temperature might rise to 140°.

If we knew exactly the law which connects the daily variations of temperature with the mean temperature, we might from the preceding data assign to the daily variation its maximum value.

When, instead of comparing the extreme temperatures of a day, we compare the two extreme temperatures of the same month, the limits of variation become greater: this is a result we might have expected, since the different causes which influence the variation of temperature may be more modified in the space of a month than during an interval of twenty-four hours. On the other hand, the distance of the sun from the equator increases or decreases more sensibly in the former period than it does in the second.

The variation is greater still on comparing the extreme temperatures of any particular month in different years: it then becomes nearly equal during each of the months of the year. It is even found that the greatest variation observed (as should be the case) is in January, the coldest month of the year.

The monthly variations no longer depend then on the elevation of temperature, nor on the length of the day. These two causes, so active on the daily period, are modified by other causes at least as influential. These seem to refer to the state of the atmosphere, which allows of radiations and chills (sometimes considerable) during the winter nights.

I have now said enough of temperatures. From these remarks, it may be easily perceived how important a part the Theory of Means should play in the discussion of observations of all kinds.

LETTER XIV.

THE LAW OF THE OCCURRENCE OF TWO KINDS OF SIMPLE EVENTS, THE CHANCES OF WHICH ARE EQUAL AND THE COMBINATIONS FEW. — THE ACCORDANCE OF THEORY WITH EXPERIENCE.

THE consideration of limits, and of the variations to which individual appreciations between such limits are subject, forms the object of a theory as interesting as it is instructive. I will not yet however touch upon this subject, but will first examine a question which immediately relates to it, and which will render the course we must pursue much more easy. It relates to the *determination of the law of the occurrence of two kinds of events, the chances of which are perfectly equal,* and which may happen either separately or simultaneously, but in different combinations.

Thus I will suppose a country in which exactly as many men die as women. If I open one of its mortuary registers, it is evident that the probability of finding a male death on the first line is equal to the probability of finding a female death. Further, if I go through the register, and count separately the deaths of males and females, theory teaches that the numbers should correspond. Perhaps experiment will not completely confirm the predictions of theory; but at least, as we have already seen, the differences will diminish in proportion as the number of recorded deaths increases.

So as not to have two things under consideration at the same time, I will content myself with stating, in the successive hypotheses I am about to make, the results of *theory,* leaving it to *practice* to corroborate them by a sufficient number of observations. If, instead of considering all the deaths inscribed, one by one, I take them *two by two,*—the first with the second, the third with the fourth, and so on in the order of inscription,—the groups may be of four kinds, —

A male, and a male.

A male, and a female.

A female, and a male.

A female, and a female.

But each of these groups forms a compound event, which has the same probability of occurrence, that is $\frac{1}{4}$. Each group then, after a sufficiently extended trial, must have occurred the same number of times. If we consider that the second and third groups are both composed of the deaths of one male and of one female, we see that the four groups reduce themselves in reality to three,—

Two males.

One male, and one female.

Two females.

Thus the second group, after the reduction, will be produced as often as the first and second combined. There is then a certain order which will be established in the *succession of events*.

Now let us take the entries *three by three*, classing together the first, second, and third, the fourth, fifth, sixth, and so on. The following will be the possible groups:—

Three males.

Two males, and one female.

One male, and two females.

Three females.

The chances in favour of each of these groups, or compound events, will be respectively 1, 3, 3, and 1; so that, after having completed the extracts from the register of deaths, and having taken each time a group of three consecutive entries, we should find that, of eight of these groups, there would be *one* of *three males*, one of *three females*, three of *two males and one female*, and *three* of *two females and one male*. Successively dividing the deaths, in like manner, into groups of 4, 5, 6, &c., we should always find the same regularity, supposing always that the number of deaths is sufficient to eliminate the effects of accidental causes.

Thus even in registers, where the deaths are apparently entered in an extravagantly capricious order, we find an admirably regular succession, whether the entries be taken singly, or by twos, threes, or any other group we please. Such a regularity is well calculated to make us reflect on what is commonly called "chance."

The following table shows the ratios in which the male and female deaths are distributed for the thirteen most simple groups possible. This table is a reproduction, under another form, of the famous arithmetical triangle of Pascal.

No. of Deaths in each Group.	MANNER IN WHICH THE DEATHS ARE DISTRIBUTED IN EACH GROUP.														Total Deaths.
	NUMBER OF FEMALE DEATHS.														
	0	1	2	3	4	5	6	7	8	9	10	11	12	13	
1	1	1	...	...	...	...	...	...	...	...	...	...	...	...	2
2	1	2	1	...	...	...	...	...	...	...	...	...	...	...	4
3	1	3	3	1	...	...	...	...	...	...	...	...	...	...	8
4	1	4	6	4	1	...	...	...	...	...	...	...	...	...	16
5	1	5	10	10	5	1	...	...	...	...	...	...	...	...	32
6	1	6	15	20	15	6	1	...	...	...	...	...	...	...	64
7	1	7	21	35	35	21	7	1	...	...	...	...	...	...	128
8	1	8	28	56	70	56	28	8	1	...	...	...	...	...	256
9	1	9	36	84	126	126	84	36	9	1	...	...	...	...	512
10	1	10	45	120	210	252	210	120	45	10	1	...	...	...	1024
11	1	11	55	165	330	462	462	330	165	55	11	1	...	...	2048
12	1	12	66	220	495	792	924	792	495	220	66	12	1	...	4096
13	1	13	78	286	715	1287	1716	1716	1287	715	286	78	13	1	8192

The numbers contained in each horizontal column express the relative chances of having in each group a fixed number of males and females. Thus the last horizontal column shows that, grouping *thirteen by thirteen*, out of 8,192 groups, there will only be 1 composed of 13 males, 13 composed of 12 males and 1 female,

78 composed of 11 males and 2 females, 286 of 10 males and 3 females, and so on. The group which combines the greatest number of chances in its favour is that which presents as many male deaths as female.

From the preceding table we may draw by induction the following conclusions, which might also be easily demonstrated by theory :—

1st. In proportion as the group of deaths is increased by unity, the total number of possible chances is doubled.

2nd. Of the whole number of chances, there is never more than one of having all male or all female deaths.

3rd. The number of kinds of different groups exceeds by unity the number of deaths forming each group.

4th. Each group has its own particular probability; and the most probable group is that which would give as many female as male deaths, when the number is even. When the number of deaths forming a group is odd, there are two kinds of groups equally possible, which have the greatest probability, which are the succession of as many male deaths as female deaths less one, or as many female deaths as male deaths less one.

5th. The other kinds of groups range on the two sides of the foregoing in a symmetrical manner; and as they become more distant, their respective probabilities diminish. It is sufficient to have written the first half of one of the horizontal columns to know how the other terms proceed.

Our numerical table contains then a scale expressing the degree or the *law* of the probability of all the compound events, so long as they do not depend on more than thirteen combinations. When the number is more considerable we may make use of a table which I shall give hereafter.

I have been showing what theory teaches on the subject of the law of the occurrence of two events which have exactly the same probability. You may be curious to know how far a sufficiently long experience would justify the prediction of calculation. I have experienced the same curiosity; but I should have renounced the pleasure of satisfying myself, were I compelled to search the mortuary registers of a town. In addition to the tedium of such a proceeding, the object on which the mind would have to dwell is not of a pleasing nature, nor would it furnish a sufficient recompense for the labour it would impose.

Happily I can resort to experiments more expeditious, and quite as conclusive. I caused to be placed in an urn 40 white balls and 40 black, so that the probability of drawing a ball of one or the other colour was the same. Then, in order that the probability might not be altered in the course of the trials, each ball drawn was replaced in the urn, care having been taken to note its draw and its colour. Using these several precautions, I caused 4,096 successive drawings to be made. Let us see what have been the results of these experiments.

First, considering the balls one by one, I found that 2,066 white and 2,030 black balls were drawn. Theory shows each of these two numbers should be equal, i. e. 2,048. If the theory does not exactly agree with experience, it will at least be admitted that the difference is but small. In fact, of 2,000 balls drawn, instead of having 1,000 white and 1,000 black, I have had on the average 1,008 white and 992 black, the difference being 8 in 1,000, or proportionally ·008,—a very small number. Let us not, moreover, lose sight of the fact that, had I had patience to make a more considerable number of trials, the difference would have been smaller still.

Afterwards, taking on my list the balls two and two,—the first with the second, the third with the fourth, and so on,—I have been able to form, by means of my 4,096 drawings, 2,048 binary groups.

The table given above shows that, for *one* group of *two white balls*, I should find *two* of *one white and one black*, and *one* of *two black balls*. Instead of which I have found

543 groups of *two white balls*,
980   „   of *one white and one black*,
525   „   of *two black*,

or rather, taking proportional numbers, which will be more useful in establishing our approximations, I have obtained 1·06, 1·91, 1·02, instead of 1, 2, and 1. The greatest difference is here ·09 in defect, which surpasses that which we have previously found, without however attaining a tenth.

I have been able to form 1,365 ternary groups by means of my

4,096 drawings. They are distributed proportionally in the following manner:—

	1·08	3·03	2·77	1·10
instead of	1·00	3·00	3·00	1·00

The greatest difference between experience and theory is here ·23 in defect, and falls on the number 2·77. Of my 1,365 groups of three balls, three-eighths (that is 510) should have included *two black balls* and *one white*,—instead of this number, I have only 474.

In proportion as the number of balls in a group are increased, the number of groups on which the experiments are made decrease.

When I have taken the balls four and four together in the order of drawing, I have formed 1,024 quaternary groups, which are divided proportionally in the following manner:—

	1·07	4·18	5·78	3·84	1·11
instead of	1·00	4·00	6·00	4·00	1·00

The greatest difference is ·22 in defect.

Of 819 groups of five balls taken at a time the proportional numbers were

	1·2	4·9	10·8	8·7	5·3	1·1
instead of	1·0	5·0	10·0	10·0	5·0	1·0

The greatest difference is 1·3 in defect. Already the differences which had successively been in third, second, and first places of decimals affect the whole numbers.

Of 683 groups of six balls at a time the proportional numbers have been

	1·6	6·1	15·5	18·0	15·5	6·5	0·8
instead of	1·0	6·0	15·0	20·0	15·0	6·0	1·0

The greatest difference is 2 in defect.

Of 385 groups of seven balls at a time the proportional numbers have been

	2·0	7·4	22·7	33·0	32·4	20·8	8·8	0·9
instead of	1·0	7·0	21·0	35·0	35·0	21·0	7·0	1·0

The greatest difference is 2·6 in defect.

Without carrying the comparisons further, it will be seen that the differences between experience and calculation increase as the number of groups on which the trials are made diminish, which is

in conformity with theory. We know, in fact, that the precision
of the results increases as the square root of a number of observations.

In the last example we have really but 385 observations to
verify 128 possible events, which scarcely gives three observations for one event. In the first example, on the contrary, we
have 4,096 observations to verify but two simple events, which is
2,048 observations for each of these simple events. But the square
roots of the numbers 3 and 2,048 are nearly as 1 to 26: the degrees
of precision ought only to be in the ratio of these numbers.

It results then, that continuing the trials to infinity, and taking
the balls either one by one, two by two, three by three, &c., we
should be certain that the drawings would present themselves as
calculation has pointed out. But when we proceed by the way of
experiment, it is impossible to make an infinite number of trials:
we should not therefore expect to obtain results agreeing exactly
with those which theory gives; or even if the agreement were
established, it might be but accidental. It, nevertheless, occurs
that errors or differences become smaller in proportion as the trials
are multiplied. Nay further, we shall be able to see that we may
assign the probability that these differences do not extend beyond
certain given limits.

The powers of man are limited. Nature is unbounded. The
Supreme Being alone can see events proceed in accordance with
his laws. To him time is nothing, and all imaginable combinations
may be realized in succession. These apparent differences are only
found within the sphere of man, and spread a remarkable variety
over all the events in which he is concerned. This variety, which
is in part his work, has however narrow limits, and cannot alter
the general order of things.

LETTER XV.

ON THE OCCURRENCE OF TWO KINDS OF SIMPLE EVENTS, OF WHICH THE CHANCES
ARE EQUAL, AND WHICH MAY BE COMBINED IN A CONSIDERABLE NUMBER OF
WAYS.—SCALE OF POSSIBILITY.—ITS CONSTRUCTION.

My preceding letter illustrates this curious fact, that if an urn contains an infinite number of white and black balls, in equal proportions, by inscribing at each drawing the ball drawn, and indicating its colour, the series of balls will proceed in the most regular manner. Thus, taking them one by one, we should find the number of white balls drawn exactly equal to the number of black balls.

Taking them two by two, the number of groups of one white ball and one black will be double the number of groups containing either two white or two black.

I have indicated the law of the drawings on the hypothesis of grouping the balls by three, four, five, up to thirteen. I have next placed the results of experiment by the side of those of calculation; and it is easy to recognise that the slight differences only proceeded from the smallness of the number of drawings which had been made. If these drawings could have been prolonged to infinity, the results observed would have absolutely conformed with those calculated.

It must not be supposed that this agreement would stop with the most simple groups. I could have extended my table further, and have assigned in advance, by calculation, the order of the drawing of balls taking them by fourteen, fifteen, or by any other superior combination,—by the thousand, ten thousand, or the million, or even taking them in a number still more considerable.

When the number of balls taken at a time is rather large, all

F

the possible chances are so multiplied that we might despair of being able to verify theory by experience. In only taking twelve balls at a time, we have already seen that at least 4,096 drawings must be made, in order that each possible group may happen once. But the number of drawings should be incomparably greater, if we desired that the groups should occur in conformity to the indications of theory.

If we took 1,000 balls at a time, in order to draw out at least *once* each possible group, with all the arrangements of which it is capable, we should require more ages than have transpired since the Creation. How long a time then would be necessary, were it wished, to repeat all the drawings possible a sufficient number of times to make experience accord with theory? It would nearly require eternity to arrive at a satisfactory result.

It is useless to dream of such proofs. Happily we can pass them by, in the applications which we shall have to make of the Theory of Probabilities to the sciences of observation.

I even hope to show that it is sufficient to have carefully calculated all the possible drawings, with the chances they present. To take a particular case. Suppose, for example, that we take 999 balls at a time; and this calculation will supply those which I ought to have made for every other combination, provided it includes a greater number of balls than that indicated in the preceding table, even if the number of the balls were infinite.

Let us then stop at the particular case of drawing 999 balls at a time. To follow the order of our first table, the event which will first present itself will be that which consists of drawing 999 white balls without the mixture of a single black ball. But this drawing will be extremely improbable: there would be but one chance for this drawing, out of a total number of chances which it would require more than *three hundred figures* to indicate.

The following event, which would consist of taking 998 white balls and one black, would also have an extraordinary small probability; and it would be the same with all the drawings of a great number of white balls with a few black balls, or a great number of black balls with a few white. These drawings might be considered as nearly impossible.

We only obtain a probability slightly appreciable by supposing that, of 999 balls drawn, there are at least 420 white, and not more than 579; the same with the black balls. In either case, the probability of the drawing is only represented by the very small fraction ·00000004.

The probability increases in proportion as the number of white balls in each drawing approaches to equality with the number of black balls; and the drawing which has the greatest probability is that of 500 white balls and 499 black, or *vice versâ*. This probability would be ·025225.

If we wished then to test theory by experiment, we should be almost certain beforehand that every draw of 999 balls would bring out nearly as many white balls as black. All the drawings, in fact, in which the balls of one colour would exceed those of the other colour by 160, would not together have *half a millionth* of probability in their favour.

It would be superfluous, in constructing a table of the possibility of drawing any group of 999 balls, to take any account of these latter drawings, the number of which mounts up to 840: we may content ourselves with calculating the 160 groups, which do not contain less than 420, and not more than 579 white balls. A table of these, called *the scale of possibility*, will be found in the notes at the end of this volume.

I dare not flatter myself that I have been sufficiently clear. If my explanations leave something yet to be desired, for want of words, perhaps I may succeed better in addressing the eye. One of the most spiritual thinkers of antiquity has said, that what we hear excites the mind less than what we see,—

> "*Segnius irritant animos demissa per aurem*
> *Quam quæ sunt oculis subjecta fidelibus.*"

Besides I have not forgotten your Highness's taste for Geometry, and particularly that which strengthens and relieves thought. Graphic constructions have been much abused, but I hope a like reproach will not be applicable to mine.

To return to the example I have chosen. Let us first recall to our minds that the drawing of 499 white balls and 500 black is that which has the greatest number of chances in its favour. The drawing of 498 white balls and 501 black has a less probability.

In proportion as the difference between the number of black and white balls in each drawing of 999 balls increases, the probability of the drawing diminishes. This decrease is shown in the figure by the heights of the small successive rectangles constructed on *a b′*, *b′ c′*, *c′ d′*, &c. Thus the probability of drawing 480 white balls out of 999 is represented by the small rectangle constructed on *o′ p′*. The table of possibility gives the number ·011794; and the figure shows that this probability is in fact little less than half the *maximum* probability, which consists of taking 499 white balls in one drawing,—a probability which is represented in the figure by the greatest rectangle, that constructed on *a b′*.

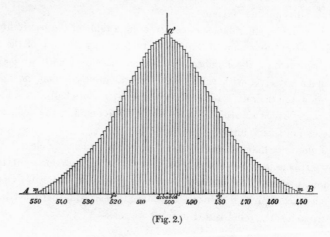

(Fig. 2.)

The small rectangles afterwards decrease so rapidly that that which represents the probability of taking 450 white balls in a drawing of 999 has scarcely any height: and yet, to follow the law of continuity, we should construct yet 450 more, each successively diminishing in height. I have not represented them in the figure, for not even a magnifying-glass would have sufficed to make them appreciable. In our scale of possibility I have been able, by means of numbers, to carry the thing a little further; but even the numbers should be neglected for the different drawings which suppose less than 420 balls out of 999.

Each of the small rectangles, in juxta-position in the figure, indicates by its size the probability of the most probable drawings of 999 balls.

The figure is symmetrical on the two sides of the axis $a\,a'$; and this ought to be the case, for what takes place for the white balls should also take place for the black. Thus the two rectangles constructed on $p\,o$ and $p'\,o'$, on the two sides of and at equal distances from the axis $a\,a'$, are equal, because the probability of taking 519 white balls and 480 black is equal to that of taking 480 white and 519 black.

A glance at the figure will make the law of the drawing of balls more evident, for the case with which we are now occupied, than all the reasoning possible.

What we might bet on each drawing individually is proportional to the height of the small rectangle which corresponds to that drawing. If the heights of these small rectangles represented piles of crowns, in collecting all those deposited between the limits m and m', I doubt not but that great embarrassment would be felt in joining to them those extremely small numbers placed without these limits: all the crowns composing them would scarcely form a thousandth part of those included within the limits stated.

LETTER XVI.

THE SCALE OF POSSIBILITY IS GENERAL IN ITS APPLICATION.

I WILL now endeavour to give more generality to the ideas thrown out in my two preceding letters, and to explain that a scale of possibility (such as I have calculated) may serve in all cases,—in those in which the number of balls shall be greater than 999, as well as in those in which it shall be less. In giving so great a generality to our table, I will sufficiently show its utility.

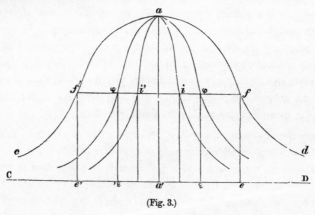

(Fig. 3.)

I suppose the curve $c f' a f d$ to represent the series of numbers contained in our table of possibility. This line is the same as that constructed in my preceding letter, only I have substituted a continuous curve for a broken line. I again stop at the points c and d, although the line should extend beyond these points to very distant limits, which I could not indicate in the diagram. We are compelled here, as with the numerical scale, to neglect that which has no appreciable value.

The line C D must be conceived to be divided into 160 equal parts, and symmetrically arranged on the two sides of the point a'. Each point of division e indicates, by its distance from the point a', the numerical difference between the balls of the two colours in a drawing of 999 balls; and the length of the corresponding perpendicular $e\ f$ represents the probability of this same drawing.

Suppose now it were wished to calculate and construct the curve of probability for a quadruple number of balls at each drawing. Take, for example, 3,996 balls instead of 999. In order to make things comparable, I will make the perpendicular $a\ a'$ again serve to represent the probability of the most probable event. Theory next indicates that the new curve of possibility will be of the form $\phi'\ a\ \phi$, in which all the perpendiculars of the first curve $f\ e$, $f\ e'$, &c., have but approached the axis $a\ a'$ in the ratio of 1 to 2; and this ratio is determined by that of the square roots of the numbers 999 and 3,996 relative to the two curves.

If we were again to quadruple the number of balls in each drawing, and made it 15,984, the curve of possibility would have the new form $i\ a\ i'$, and would be narrowed to one-half of its width with respect to the curve $\phi'\ a\ \phi$, and to one-fourth with respect to the curve $f'\ a\ f$, which means that, *cæteris paribus*, the numerical difference between the balls of the two colours would be proportionally reduced one-half with respect to drawings of 3,996 balls, and one-fourth with respect to drawings of 999 balls.

The more considerable that the number of balls taken at each drawing is, the more it tends to establish an equality between the numbers of white and black balls. If even things were carried to the extreme, and if each drawing brought out an infinite number of balls, the line of possibility would sensibly reduce itself to the perpendicular $a\ a'$,—that is to say, we should be certain of taking at each drawing of an infinite number of balls as many white as black.

The line $c\ a\ d$ is sufficient then to represent in what manner all the results group themselves round the mean, when they are sufficiently numerous.

Let us now test theory by experiment. I suppose myself before an urn containing an infinite number of white and black balls, mixed in a perfectly equal proportion. The contents are concealed from me; but I am allowed to make trials or successive drawings to ascertain what the urn contains. Let us see the course I have to pursue.

If I am only permitted to make one drawing, must I, to attain with certainty my end, take few or many balls? Good sense tells me, before theory, to take a large number. In fact, if I took but one, it would be impossible for me to know how the balls are distributed, since I should only have taken from the urn one black or one white ball.

If I take six, for example, it may present seven different cases: I may either take

Six white.

Five white and one black.

Four white and two black.

Three white and three black.

Two white and four black.

One white and five black.

Six black.

But the seven drawings are not equally probable: the first has less chance than the second,—the second less than the third; and that which has the greatest chance is the drawing of three white and three black. We have for this drawing twenty out of sixty-four possible chances, as is shown by the table already given in my fourteenth letter.

We have then some prospect of making this drawing. If we miss it, we may perhaps take two white and four black balls, or four white and two black, which would give us an approximate idea of the general distribution of balls in the urn without showing it exactly. But we have a great probability of making one of these three drawings; in fact, the last two drawings have each 15 chances in their favour. Uniting these to the 20 chances we already have, we shall have 50 chances out of 64 of taking two, three, or four white balls, among six balls drawn.

Let us, moreover, remark that it is always to approximations, more or less great, that we should confine ourselves in the sciences of observation. When we light upon the true result, it is but by accident.

Let us now suppose that we pass on to large numbers, and take 999 balls at a time. We are nearly sure that we could not take balls all of one colour. It is even almost impossible not to have from 420 to 579 white balls among 999 balls drawn, as may be seen in my fifteenth letter. But, supposing we were unfortunate enough to take one of these numbers, it must, nevertheless, be agreed that we should not be far from the truth, if we concluded, after the single trial made, that the urn contained 420 white balls out of 999; but, more generally, if we did not take exactly 499 or 500 white balls, we should differ very little from these numbers.

The probability of each drawing is indicated in our scale of possibility; and in order to admit of more easy comparison, I have represented by 1, in the table annexed,* the probability of taking as many white as black balls, and I have expressed in parts of this unit the probabilities of the other drawings.

The probability is reduced to a half when reference is made to drawing 481 white balls of 999 taken. It is reduced to a thousandth for 441 white balls, and is no more than ·000003 for 420. So that we might bet a million against three that, in taking 999 balls, we should sooner fall upon the true ratio than on a combination including 420 white balls. In all cases, we are nearly certain that the drawing we make will be one of those included in our scale, and will consequently contain a number of white balls not less than 420, nor more than 579. This error, whether in excess or defect, is 79 out of 999, or nearly eight white balls out of every hundred drawn.

Let us proceed further still; and instead of 999 balls, let us take 9,999. For experiment to agree with theory, we should at each drawing take 4,999 or 5,000 white balls. If this drawing is little probable, from the great number of possible combinations, at least we shall be nearly certain not to differ much from it. We already know that the limits become more narrow in proportion as the square root of the number of observations increases. This should be understood in the following manner:—

* *See* Notes.

Previously we supposed a drawing of 999 balls,—we now suppose 9,999. But the square roots of these numbers are as 1 to 3·18 nearly: it is in this ratio also that the limits narrow themselves. When we took but 999 balls, we were convinced that in our drawing there would be from 420 to 579 white balls,—that is to say, that we should not err eight balls per cent. more or less. Well, we may assure ourselves with the same probability, when we take 9,999 balls, that the drawing exposes us to an error at the least three times smaller, and that we shall not err three balls per cent. in excess or defect.

In taking 99,999 balls at a time, we may affirm with the same probability that the balls we have drawn are divided in such a manner that, out of 100, the error we have to apprehend does not amount to unity. If, consequently, the urn contains as many white balls as black, we should have taken 49, 50, or 51 per cent., and all other combinations may be looked upon as nearly impossible.

Your Highness will judge with what a great probability it might be affirmed that an urn contains equal proportions of white and black balls, if experiment had given this equality by a drawing comprising an exceedingly great number of balls.

These results may be of useful application in the physical and chemical sciences. If we operate on a large number, and if, instead of equality, we obtain a difference, there should be some cause for it, either that the equality does not exist, or that we have not operated correctly.

Let us now return to the example chosen. We have taken a certain number of balls,—999 for instance. Of this number, how many white and how many black do we hold? We can affirm nothing; but we can assign the probability which belongs to each possible drawing. But, as we have seen, each of the probabilities is already calculated in our scale of possibility; only we have omitted to give the probabilities for those drawings which are almost impossible.

Now the experiment is made. The hand is opened; and we count the balls, separating the white from the black. Shall we actually have as many white balls as black? Probably not; but we shall have some approximate combination, and so much the nearer as our drawing contains a larger number of balls.

A chemist would have the greatest chance of arriving at results in conformity with truth, were he perfectly sure of his analyses. In drawing a glass of water from a pure source, the atoms of oxygen and hydrogen would be found in an infinite number, and consequently in the ratio fixed by the Creator in the composition of water. This case corresponds with the extraction of an infinite number of balls from the urn.

LETTER XVII.

THEORY OF MEANS.

I HAVE shown in my three preceding letters the law of occurrence of an event, the number of chances in favour of which is equal to the number of chances against it. I hope I may have been clearly understood, and that the details into which I have been compelled to enter have not been destitute of interest.

I have particularly insisted on the case where the chances were 999 in number, and could, by their combinations, give rise to 1,000 different events. I will now endeavour to illustrate how this theory is directly connected with the Theory of *Means*.

I will take an example in which the results of observations may differ one from the other by a thousand degrees of magnitude. For this purpose, I will suppose that I have measured the height of the Drachenfels, whose summit commands so majestically the course of the Rhine, and seems to bid a last adieu to that fair stream at the moment of its quitting the valley in which it had been embosomed. I will further suppose that, instead of obtaining the actual height, I have each time found different values, but without any particular cause being in existence which would involve a liability to error either in excess or in defect. The mean of all these separate values is a number which is not probably the real height of the mountain, but which does not however materially differ from it. I do not know how far the limits of errors might extend, were I to continue my observations. I substitute then, for an unknown quantity, some known quantity. What is required is to ascertain the amount of error to which I am liable.

Considering the matter in a general way, the true height of the mountain corresponds to the drawing an equal number of white and black balls,—that is to say, that among all the possible combinations, there would be found in the long run as many measurements in excess as there are in defect of the true height sought, which would be equal to their mean. Instead of this value, I find another, which will approach that which I should have in proportion as (all things being equal) the number of my observations is increased. So much for the precision of my final result.

But, beside the question which we have just considered, another is presented, perhaps more curious still. It is this,—the results observed do not range indifferently on either side of the mean, but *in a determinate order*, which is that assigned by the scale of possibility.

Thus, when a height is measured, between the greatest and least values obtained there are presented others at different distances from the mean, but the greatest number occur in the immediate neighbourhood of this mean. This arises from the fact of the probability of the result diminishing more and more according as the measurement differs from the true number sought, in conformity with the law which regulates the occurrence of two kinds of errors the chances of which are equal, and which may be combined in different manners. Supposing that we have measured the Drachenfels so frequently as to be justified in the belief that the difference between theory and calculation has disappeared, the mean of all these measurements would give the height sought. Further, grouping the observations in order of magnitude, and proceeding by differences of 4 inches, the two greatest groups will be those which would give the height of the mountain with an error of 4 inches more or less. The two following groups will give the height with an error of 8 inches, and so on. The further we depart from the mean, the fewer observations will each group include.

We are now in a condition to recognise, *à priori*, the order which will be established among the errors. Each series of observations has its particular scale of possibility, as I have indicated in my preceding letter. The nature of this scale is determined by the number of observations, as also by the more or less precise means employed in making such observations. It must not, moreover, be forgotten that we always rigorously proceed on the hypothesis that the chances of error in excess are the same as those of error in defect. Is it not marvellous, that the errors we make accidentally range themselves in an order so perfect? and that mistakes proceed, without our knowledge, with a symmetry which seems to be the result of the best reasoned combinations?

The curve which traces the order of succession, and the number of the results, will be identical with the curve of possibility, inasmuch as an error more or less great should be assimilated to the drawing of a number of black and white balls, differing more or less from equality in regard to numbers.

If the chances of error in excess were not the same as those of error in defect, the curve tracing the succession of the results would lose its symmetry, without ceasing to exhibit still a certain degree of regularity: its summit would no longer be at an equal distance from its two extremities. I shall have occasion to revert to this curve, when I shall hereafter notice in a general manner the influence of accidental causes.

When we content ourselves with taking an arithmetical mean of different quantities, entirely independent of each other, the line which traces the observations no longer presents any kind of regularity; or if there be any, it can be but accidental.

Compared with the theory of balances or of levers in mechanics, the Theory of Means offers curious analogies. I content myself with citing the principal: it may contribute to the casting of a new light on this interesting subject.

Many parts have been measured off from the point O, along the right line O B.

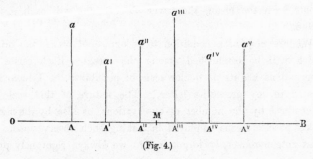

(Fig. 4.)

The part O A has been measured twenty times, and I indicate it by placing on its extremity A a pile A a of twenty shilling pieces. The part O A′ has been measured eleven times: I indicate it also by placing on its extremity A′ a pile of eleven shillings. I do the same for each of the other parts measured; so that each pile or each perpendicular, A a, A′ $a′$, A″ $a″$, marks by its height the number of observations relative to each length measured.

We now conceive that all these piles of shillings form together a certain weight; and if they are placed on the beam of a balance, or on a lever O B (which we will suppose, for the sake of simplicity, to be without weight), in order to sustain them *in equilibrio* on this lever, a convenient point of support must be selected: this point will be exactly at M, at the extremity of O M, the mean of all the lengths O A, O A′, O A″, &c. The piles of shillings on the right of the point M will balance those on the left. The result would be the same as though as many shillings were piled on the point M as are distributed along the lever.

It is sometimes useful to represent a succession of observations by means of a curve. A simple inspection of such curve may indicate whether there exist a law which connects numbers, which we might have thought to be absolutely independent of each other. I shall have occasion to cite many curious examples. There is one in particular, to which I attach the greatest importance: it relates to the theory of man, and may in some degree serve as a base to this theory, considered in a scientific point of view.

Notwithstanding my desire to close the somewhat abstract discussion which has occupied us some time, I cannot conclude without adding a few words on the manner of measuring the degree of precision of a series of observations. This will form the subject of my next letter, which I promise shall be as short as possible.

LETTER XVIII.

I STATED, at the conclusion of my last letter, that we have from this time a mode of measuring the degree of precision of a series of observations. I will even add, that it is possible to assign the relative degrees of precision of many series of different observations. Thus two persons have measured 1,000 times in succession the height of the same edifice; and it is wished to know what are the degrees of confidence to be attached to their results.

If these two observers have no tendency to observe too high or too low, and if their errors are but accidental, we already know that the two series of observations may be represented by two curves of very irregular form.

I have remarked, as a result of theory, that supposing the observations to be *equally* well made and equally numerous, the two curves which represent their succession should be such as could be placed one on the other; and that if the observations be unequal in number, one curve would be more narrow than the other. The contraction towards the axis is proportional to the square root of the number of observations.

I will now add, (and it is the principal object of this letter,) that supposing the observations equally numerous in each series, but *not equally* well made, one of the curves will be more contracted than the other. *The contraction towards the axis is proportional to the degree of precision of the observers, and gives the measure of this precision.*

Our aim then should be to seek means of appreciating the contraction of the curve. I have calculated a table to this effect, and I have named it *the scale of precision:* I have deduced it, without trouble, from the table of *possibility.*

I first take the first number of this latter table, which also becomes the first number in the table of precision: I then add the
second number of the table of possibility: then to this sum I add
the third number of the table of possibility; and I continue to add
successively in this manner all the numbers contained in my first
table, placing the several sums in my second table.

Proceeding thus by addition to the end of the table, I only obtain the probabilities of one-half of all the possible drawings of
those in which the white balls are inferior in number to 500. The
sums should be doubled to obtain the probabilities of the drawings
in the other category. Some examples will render this more intelligible.

The probability of taking more than 480, and less than 519
white balls, is composed of the sum of the probabilities of taking
481, 482 . . . . 518 white balls, which will give the double of
0·397172, or nearly ·8. This is the probability of not differing
from the mean, either in excess or defect, by ·020 of the extent of
the limits between which the combinations may vary. In the
figure which represents the curve of possibility this probability is
represented by the surface, $f\,e$, $f'\,e'$, of the forty contiguous rectangles constructed on $f'\,i\,f$ symmetrically on the two sides of $a\,a'$.

The probability of not differing from the mean in excess or
defect by ·01 of the extent of the limits is the double of ·236548,
or nearly one-half. We may then bet 1 to 1 that such a difference
will not occur. This difference of ·01 is very frequently employed,
—it has received the name of *probable error:* it serves to measure
the degree of contraction of the curve of possibility, and consequently becomes the modulus of precision.

I now come back to the example mentioned at the commencement of this letter; and I suppose two persons who, with a different degree of address, have each measured 1,000 times in succession the same edifice. The two curves which will represent
the observations will be unequally contracted towards the mean.
Let us see how it will be possible to appreciate the degree of contraction, which will serve as measure of the precision.

I will commence by grouping the two series of observations,
proceeding by differences of $\frac{1}{4}$ of an inch. The most numerous

groups will be those which are most approximate to the mean, as we have already seen. I make then the sum of the observations contained in each group, taken in each direction from the mean; and I stop the addition when the sum is $\frac{1}{2}$. I note with care how far I should err from the mean to arrive at this sum. I will suppose that it is at distances of $+ 1\frac{1}{2}$ inches and $- 1\frac{1}{2}$ inches. These are the terms which are called the limits of probable error. There are as many observations within these limits as beyond. We might bet then 1 to 1 that the same observer, if he recommenced his work, would make errors which would be within these limits. It is the greatness of the probable error which will henceforth serve us for *a modulus of precision.*

Suppose that we operate in the same manner for the second series of observations, and that we find as the limits of probable error $+ 2$ inches and $- 2$ inches: we may say that the relative precisions are as 6 to 8.

In the preceding example I have passed over the scale of precision, because I have supposed that, by adding the groups of observations which differ the least from the mean, I have obtained a sum forming exactly $\frac{1}{2}$. But this but seldom happens, and in the greater number of cases recourse must be had to the scale of precision.

An example presented with a few details would easily explain the course to pursue; but I have not forgotten that I have promised to be brief: I therefore give this example in another letter, which your Highness may pass over, if in haste to arrive at the applications which I purpose making of the Theory of Means to the knowledge of man.

LETTER XIX.

THE example which I am about to present is taken from one of those sciences which proceed with the greatest precision in their researches. Astronomy makes a frequent use of means. The excellence of the instruments, and the certainty of the methods it employs, induce us to think that it yields in point of exactness to no other science, and that consequently the variation of the numbers from the mean are always included within very narrow limits.

An observer may assign the position of a star without fear of errors of more than three or four seconds in arc,—that is to say, that the extent of his error will at most be but the breadth of the small band of the heavens, which would be concealed by a thread placed at the distance of several feet before our eyes. However, a great number of causes may give rise to this error, and we here place ourselves in the most unfavourable hypothesis : we suppose that they all tend in the same direction to give a value either too great or too small. Thus, however precise the instrument may be, it is not perfect in all its parts,—whatever may be the skill and experiènce of the observer, his sight is not infallible : the state of the atmosphere may be more or less unfavourable. We only see the stars through the atmosphere in which we are placed, and, by reason of refractions, they are not really in the places where we perceive them. Science, it is true, has given us means of estimating these displacements, and of correcting them, but in a manner the less precise the nearer we approach the horizon. Notwithstanding these causes of error, and many others which it is superfluous to mention here, a practised astronomer, using a good instrument, may (as we have already said) fix by a single observation the position of a star within three or four seconds; and we then admit the concurrence of the most unfavourable circumstances,—those where all the causes of error act in the same direction. Rarely have we to fear such errors; but rarely also do we obtain exactly the number to be determined.

I believe it is superfluous to mention that, in the example with which we are about to be occupied, I continue to admit that the chances of making errors in excess are exactly the same as those of making them in defect. The observations which I shall use are taken from the publications of the Royal Observatory at Greenwich: they refer to the determinations in time of the right ascension of the Polar Star; and by these words "right ascension" must be understood the distance of a star from the equinoctial point, measured along the celestial equator. I have employed for this purpose the observations of the Polar Star, obtained during the four years 1836 to 1839 inclusive. These observations have been corrected for nutation, precession, &c., and have been calculated for one particular time; so that they may not differ from one another, but by the effect of small accidental errors.

The first column of the following table indicates how much the observations differ from the mean in excess or defect. In order to be able to group the observations, I have marked $0^s \cdot 5$ for the difference in right ascension where the variations from the mean were included between the limits $0^s \cdot 25$ and $0^s \cdot 75$. In the same manner, the differences which fall between $0^s \cdot 75$ and $1^s \cdot 25$ have been united as forming variations of one second; and so on. As to the observations which do not differ $0^s \cdot 25$ from the mean more or less, they are considered as equal to the mean.

We remark that the different groups given in the second column are nearly such as they should be, according to theory,—that is to say, that which includes the greatest number of observations is at the mean. The two next largest groups are those which fall near the mean, and which correspond to the two differences $+ 0^s \cdot 5$ and $- 0^s \cdot 5$. The groups then continue to decrease as they become more distant from the mean.

Difference in Right Ascension by Variations from the Mean.	Number of Observations.		Probability of the Variations according to Observation.	Rank of the preceding Numbers in the Scale of Precision.
	Absolute.	Relative.		
— 3^s·5	1	2	·500	
— 3·0	6	12	·498	45·5
— 2·5	12	25	·486	35·0
— 2·0	21	43	·461	28·0
— 1·5	36	74	·418	22·0
— 1·0	61	126	·544	16·6
— 0·5	73	150	·218	9·3
			·68	2·6
MEAN	82	168	·100	4·0
+ 0·5	72	148	·248	10·5
+ 1·0	63	129	·377	18·5
+ 1·5	38	78	·455	27·0
+ 2·0	16	33	·488	35·5
+ 2·5	5	10	·498	45·5
+ 3·0	1	2	·500	
	487	1000		

The number of observations is 487: but if we take the three greatest groups, including 82, 73, and 72 observations, 227 in all, we shall have little less than half the sum 487; and consequently the probable error will be little more than 0^s·5, as may be seen from my preceding letter. I am about to seek a greater precision for these different results.

I will first substitute for the absolute numbers of observations, written in the second column, the relative numbers if 1,000 observations had been made; so that each number in the third column represents, according to experiment, the number of chances out of 1,000, or the probability which each corresponding variation in the first column has in its favour. Thus 0·148 is the probability indicated by the observations of an error from the mean equal to $0^s\cdot5$, if the observations were made with the same instruments and under the same circumstances.

The third column would present then, had the experiments been sufficiently prolonged, a true scale of precision, applicable to the observations that might afterwards be made at Greenwich, with the same means of observation. By the aid of this scale I have calculated the scale of precision contained in the fourth column.

The fourth column has been formed by means of the third, in the following manner. From the number 500, written at the foot of this last column, I have deducted the number 2 placed opposite in the third column; then, from the remaining 498, I have deducted the number 10, also placed opposite in the third column; and so on, up to a remainder of 100. This remainder expresses a part of the observations which enter the group of the mean. It should, strictly, be equal to one-half of 168, because the group of the mean contains the observations in which the error does not exceed $0^s\cdot25$ in excess or defect; and the two portions of the group should be theoretically equal. We also perceive that the general mean has not been altogether well calculated, and that it should fall a little lower. I have next made a calculation analogous to the preceding, commencing at the other end of the column. This time I again have a remainder of 68, which, added to the 100 obtained already, reproduces 168.

When the fourth column has been formed in this manner, it shows, according to experience, the different probabilities that the errors to which we are exposed will not vary from the mean beyond the limits indicated in the first column. Thus there are 418 chances out of 1,000, or 0·418 as the probability that we shall not vary $1^s\cdot5$ from the mean in defect; and in the same manner 0·455 is the probability that we shall not differ $1^s\cdot5$ from the mean in excess.

Strictly, these two probabilities, 0·418 and 0·455, should be

equal. But it may be conceived that the probabilities deduced
à posteriori are so much less exact as the observations from
whence they are deduced are the fewer in number.

I shall henceforth suppose that, when we have a series of observations, we always commence by calculating a table analogous to
the preceding: we shall thus have, in the third and fourth columns,
scales of possibility and of precision relative to the observations
with which we are about to be occupied.

The last column of the table gives a first example of the use that
can be made of the numbers contained in the preceding column.
These numbers have been compared successively with those in our
general scale of precision, to ascertain the places they there occupy.
Thus the probability 0·418 relating to the variation — 1^s·5 is found
in the twenty-second place. These last numbers occupy the fifth
column. It is here that we may see whether the observations
proceed with regularity, and whether they vary from the mean
in conformity with the indications of theory. Mathematically the
differences should be nearly equal, except for the extreme terms.
But this is what we perceive to be the case, for we may consider
the distance of one term from the following as being represented
sufficiently exact by the number 6·5.

If, in fact, we take this quantity as the difference of two arithmetical progressions, the one increasing, and the other decreasing,
having 2·5 as their first term, we shall have a series of numbers
which will represent almost exactly those contained in the first
column of the preceding table, with the exception of the extreme
terms. This may be seen in the following table, the second column
of which is composed of numbers calculated in this way.

The third column is composed of numbers taken from the general table of precision, in the ranks indicated by the numbers of
the second column: it shows the calculated probabilities of the
different variations.

The second and third columns only reproduce in some measure
the fourth and fifth columns of the preceding table; only I have
slightly altered the observed numbers, to subject them by calculation to the law of continuity indicated by theory. The third
column shows that 496 of these numbers are less than the mean

by 3^s. There were then four smaller still: this is indicated by the number at the head of the fourth column with respect to the variation of $3^s\cdot5$: 486 observations have given numbers less than the mean by $2^s\cdot5$. The difference between 496 and 486 represents the group of observations which differ — 3^s from the mean: this difference 10 has also been inscribed in the fourth column. All the other numbers in the fourth column are obtained in like manner, by taking the difference of two consecutive numbers in the third.

Differences in Right Ascension with regard to the Mean.	Rank according to Calculation.	Probability of Variations according to Calculation.	Number of Observations in each Group.	
			By Calculation.	By Experience.
— $3^s\cdot5$	...	...	4	2
— 3·0	41·5	·496	10	12
— 2·5	35·0	·486	22	25
— 2·0	28·5	·464	46	43
— 1·5	22·0	·418	82	74
— 1·0	15·5	·336	121	126
— 0·5	9·0	·215	152	150
	2·5	·063		
MEAN	4·0	·100	163	168
+ 0·5	10·5	·247	147	148
+ 1·0	17·0	·359	112	129
+ 1·5	25·5	·431	72	78
+ 2·0	30·0	·471	40	33
+ 2·5	36·5	·490	19	10
+ 3·0	...	...	10	2
			1000	1000

This third column, in which I have established according to the table of precision the continuity required by theory, presents numbers which differ very little from those which experience has in reality furnished. It would be difficult for calculation and experiment to agree better. This agreement is a proof of the skill of the English observers.

I wish to be allowed to make one further remark. We have just seen that a difference of $0^s{\cdot}5$ in the right ascension of the Polar Star corresponds in the table of precision to six ranks and a half: proportionately, twenty-one ranks should correspond to $1^s{\cdot}61$. But these twenty-one ranks constitute the probable error: there is then 1 to bet against 1 that the variations from the mean do not pass the half of $1^s{\cdot}61$ or $0^s{\cdot}8$. The *probable error* is then here eight-tenths of a second in time.

LETTER XX.

TO DISCOVER WHETHER THE ARITHMETICAL MEAN IS THE TRUE MEAN.-TYPE OF
THE HUMAN SIZE.

THE Gladiator is certainly one of the most beautiful works of ancient sculpture. It is with reason that artists have studied its free and noble forms, and have often measured the principal dimensions of the head and of the body to obtain its proportions and its harmony.

To measure a statue is not so easy an operation as might at first appear, particularly if it be desired to obtain very precise results. In measuring ten times in succession the circumference of the chest, we are not sure of finding two results identically the same. It almost always happens that the values obtained are more or less distant from that sought; and I even suppose the most favourable circumstances those where there is no tendency to make the measurements either too small or too great.

If we had the courage to recommence a thousand times, we should in the end have a series of numbers differing from one another, according to the degree of precision exercised in their collection. The mean of all these numbers would certainly differ very little from the true value. Moreover, in classing all the measurements in order of magnitude, we should be not a little astonished to find the groups succeed one another with the greatest regularity. The measurements which differed the least from the general mean would compose the largest group; and the other groups would be so much the smaller as they contained measurements differing the more from this same mean. If the succession of groups were traced by a line, this line would be the curve of possibility: this result might in fact have been foreseen. So that unskilfulness, or chance (if to gratify our self-love we prefer this word), proceeds with a regularity which we should have been but little inclined to have attributed to it.

I now suppose the five hundred measurements differing the least from the mean to be collected together: the half-difference between the largest and the smallest of all these measurements will be the modulus of precision, or the probable error. It might happen, in the actual circumstances, that this probable error did not amount to the twenty-fifth part of an inch; so that, out of a thousand measurements, five hundred would err the twenty-fifth part of an inch in excess, and five hundred the twenty-fifth part of an inch in defect. It would then be an even wager that, in taking another measurement, we should not differ more than a twenty-fifth part of an inch from the mean of all the measurements, which may be considered the true circumference we wish to ascertain.

If we had to measure the chest of a living person, instead of that of a statue, the chances of error would be much more numerous; and I much doubt whether, after a thousand measurements, we should still find a probable error of the twenty-fifth part of an inch. The single act of respiration, which causes each instant a variation of the form and dimension of the chest, would add a powerful cause of error to those which concur in operating on a perfectly motionless statue. Notwithstanding this disadvantage, the thousand measurements grouped in order of magnitude would yet proceed in a most regular manner. The line which represents them would always be the curve of possibility, but dilated in a horizontal direction in proportion to the probable error.

Let us modify our hypothesis still more, and let us suppose that a thousand sculptors have been employed to copy the Gladiator with the greatest care imaginable. It cannot be expected that the thousand copies made would each be an exact reproduction of the original, and that, in measuring them successively, the thousand measurements which would be obtained would agree as well as though all had been taken from the Gladiator itself. To the first chances of error would be added the inaccuracies of the copyists; so that the probable error would perhaps be very great. Notwithstanding this, if the copyists have not worked with preconceived ideas, in exaggerating or diminishing certain proportions, according to the prejudices of their particular school, and if their inaccuracies are but accidental, the thousand measurements grouped in order of magnitude would still present a remarkable regularity, and would succeed each other in the order assigned by the law of possibility.

Your Highness smiles. You will doubtless tell me that such assertions will not compromise me, since no one will be disposed to make the required experiment. And why not? I shall perhaps astonish you very much by stating that the experiment has been already made. Yes, surely, more than a thousand copies have been measured of a statue, which I do not assert to be that of the Gladiator, but which in all cases differs but little from it. These copies were even living ones, so that the measurements have been taken with all possible chances of error: I will add more, that the copies have been subject to deformity by a host of accidental causes. We ought then to expect here a very considerable probable error.

I come to the fact. We find, in the thirteenth volume of the *Edinburgh Medical Journal,* the results of 5,738 measurements of the chests of the several soldiers of the different Scotch regiments. The measurements are given in inches, and are grouped in order of magnitude, proceeding by differences of 1 inch. The smallest measurement is about 33 inches, and the greatest 48. The mean of all these measurements gives a little more than 40 inches as the average circumference of the chest of a Scotch soldier: this is also the number which corresponds to the largest group of measurements; and, as theory points out, the other groups diminish in proportion as they recede from it. The probable variation is 1·312 inches,—a value of which we should not lose sight.

I now ask if it would be exaggerating, to make an even wager that a person little practised in measuring the human body would make a mistake of an inch in measuring a chest of more than 40 inches in circumference? Well, admitting this probable error, 5,738 measurements made on one individual would certainly not group themselves with more regularity, as to the order of magnitude, than the 5,738 measurements made on the Scotch soldiers; and if the two series were given to us without their being particularly designated, we should be much embarrassed to state which series was taken from 5,738 different soldiers, and which was obtained from one individual with less skill and ruder means of appreciation.

The example which I have cited merits, I think, great attention: it shows us that the results really occur, as though the chests which have been measured had been modelled from the same type from the same individual,—an ideal one if you will, but whose proportions we ascertain by a sufficiently long trial. If such were not the law of nature, the measurements would not (spite of their imperfections) group themselves with the astonishing symmetry which the law of possibility assigns them.

Of the admirable laws which Nature attaches to the preservation of the species, I think I may put in the first rank that of maintaining the type. In my work on *La Physique Sociale*, I have already endeavoured to determine this type by the knowledge of the human mean. But if I mistake not, what experiment and reasoning had shown me, here takes the character of a mathematical truth.

The human type, for men of the same race, and of the same age, is so well established that the differences between the results of observation and of calculation, notwithstanding the numerous accidental causes which might induce or exaggerate them, scarcely exceed those which unskilfulness may produce in a series of measurements taken on one individual.

If it is objected, that those men are rejected from the regiments who are deformed either by an excess of fatness or of leanness, I reply that, admitting all, we should only enlarge the limits of probable error, without altering the law which presides over the assemblage of the numbers. I could cite examples in support of this assertion, and relate the results of measurements that I have myself made on a great number of individuals without previous selection; but I have thought it right only to employ numbers collected by the hands of others, so far as that was possible.

If there existed no law of possibility which presided over the development of man, if all were chance, I (in my turn) ask how much should we not have to wager against 1 that 5,738 measurements, taken on as many chests, would range themselves in an order quite different from that determined by the law of possibility?

LETTER XXI.

EACH RACE OF MEN HAS ITS PARTICULAR TYPE.—APPLICATION OF THIS THEORY.

SINCE writing my last letter, I have thought that a very plausible objection might be made to its contents. This objection will certainly not have escaped your Highness, accustomed as you are to the sight of regiments, and to taking notice of the size of soldiers. I might be asked what would become of my pretended regularity in the manner in which measurements proceed, if I had to operate, for example, on a regiment of cuirassiers, only containing very large and strong men, and on a regiment of chasseurs, composed of much smaller men? Mixing together all the measurements, and grouping them in order of magnitude, the most numerous group would certainly not correspond to the mean.

My answer is easy. The example taken answers to that of taking from all the measured men of a nation those who form any two groups, more or less differing from the mean size. We might in this case take an arithmetical mean; but we should not have a true mean, which can only depend on the whole of the observations. When we seek a verification of the law I have endeavoured to establish, we must not make any choice, we must take all the men of a nation as they are. Thus we should not stop to consider such and such a regiment: we should in such case have but one group of men of a certain size, among all the groups that might be made with regard to magnitude. If the law is verified in respect of the Scotch soldiers, it is because the measurements have been taken on different regiments, and because also we have operated on the circumferences of the chests,—an element on which selection has less influence than on the height. We should also take care, in order to render all chances equal, only to submit to similar tests men of the same age.

Now I have been able to make this experiment. The French conscripts are measured every year; and M. D'Hargenvilliers has shown how 100,000 conscripts are distributed in order of size, grouping them by differences of 1 inch. Adopting his numbers, and comparing them with our table of possibility, I have found that they proceed in the most regular order, and range themselves symmetrically on the two sides of the mean; in other words, the numbers group themselves with the same order as if they had been measured on the same individual 100,000 times in succession, with a probable error of 2 inches.

We must doubtless measure very unskilfully, to make it an even wager that we should be in error 2 inches in the appreciation of a height of five feet and a half. However, I ask, "Could they who, in the recruiting councils, are charged with measuring the French conscripts measure the Gladiator with a probable error of less than 2 inches?" For myself, I doubt it. In the attitude of a combatant which the figure has, it becomes very difficult to appreciate the size : in order to arrive at precision, this operation should be performed by some one acquainted with anatomy. Each part should be measured separately; and great errors would be committed unless allowance could be made for the manner in which each member is disposed. An anatomist, indeed, endowed with dexterity, and accustomed to these sorts of appreciations, might give the height of the Gladiator with nearly as great chance of exactness as if the statue were upright. But it would not be so with another man; and I think I do not exaggerate in saying that the probable error would amount under certain circumstances to 2 inches.

Thus by only bending the human body, although but in a statue, we expose him who has to measure the height to errors so great, that the numbers which will be obtained will differ quite as much from each other as the numbers we should obtain by measuring all the men of one particular age taken from one particular nation. The difference which Nature makes in the heights of men is not greater than that which inexperience would produce in the measurements taken on one individual man in an attitude more or less curved.

Everything occurs then as though there existed a type of man, from which all other men differed more or less. Nature has placed before our eyes living examples of what theory shows us. Each people presents its mean, and the different variations from this mean, in numbers which may be calculated *à priori*. This mean varies among different people, and sometimes even within the limits of a single country, where two people of different origins may be mixed together.

Let a desert island be peopled to-morrow, by placing upon it 1,000 men of the tallest race, the Patagonians for example, all six feet high, and 1,000 Laplanders only four feet and a half high, the mean height in this island will be five feet and a quarter, and yet not one man will be of that height. Grouping them in order of size, we could form but two groups, and the law of possibility will be completely in fault,—it would in appearance at least be so. But we see at once that the difference only proceeds from the mixture of heterogeneous things,—men of different races, who have different laws of development.

However, let us not reject this example,—it may be useful. Let us suppose that, instead of choosing 1,000 Patagonians of the same height, we take them as they come, some being more and some less than six feet in height: when we group them in order of size, their arrangement we already know will be determined by the law of possibility. If we had done the same with the 1,000 Laplanders, it might happen that a certain number of these latter were of the same size as the smaller Patagonians; and then the two lines which would trace their several arrangements would encroach the one on the other. This encroachment would be so much the greater as the two races of men mixed together differ the less in height, and as they shall have been the more promiscuously taken. If we had to measure the sizes of a similar people, we might be ignorant that a mixture had taken place; but experiment would make it known. The line which would represent the measurement would have two summits, which would announce two different races having unequal mean heights. The law of possibility has then this new advantage, that it assists in the resolution of a problem very interesting in anthropological respects.

The case of which I have been speaking would still occur, if we made no distinction of sexes in measuring the persons of a particular nation of the age of 30 years. The men would have a different curve of height to the women, and the common curve for both sexes would have two summits. Calculation might assist in separating what belongs to one or the other sex.

Since I have begun to cite the advantages that may be obtained by the use of the table of possibility, I will give a new example. I will suppose that we only know a few groups of the whole of a series of observations, and it is required to determine the other groups.

I will again take the already cited example of the French conscripts. M. D'Hargenvilliers, in stating how 100,000 men are distributed in order of height, and in groups differing from one another by 1 inch, only indicates the total number of conscripts less than 5 feet 1 inch, and of conscripts more than 5 feet 9 inches: there are 28,620 in the first category, and only 2,490 in the second. By means of the seven intermediate groups, I have been enabled to ascertain how to re-establish the eight groups of men comprised in the case of rejection for deficiency in height, and the seven groups of men of upwards of 5 feet 9 inches.

I have, in addition, found that the probable error was nearly 2 inches. But if we notice that the mean height was 5 feet 4 inches, we shall perceive that the probable error forms $\frac{1}{32}$ of the quantity to be measured. In the example which relates to the measurement of the chests of the Scotch soldiers, I found as the probable error 1·312 inches in a circumference of 40 inches; the probable error being in this case $\frac{1}{30}$ of the quantity measured. We may hence conclude that, relatively to the sizes, the measures have nearly the same degree of precision in each case. The law of continuity of numbers shows also that two individuals only of 100,000 in France are more than 6 feet 3 inches in height, and that four are less than 4 feet 2 inches.

This same law of continuity enables us to recognise a more remarkable fact: we might suspect it, but here we find it proved, —it is that the number rejected for deficiency in height is much exaggerated. Not only can we prove this, but we can determine

H

the extent of the fraud. The official documents would make it appear that, of 100,000 men, 28,620 are of less height than 5 feet 2 inches: calculation only gives 26,345. Is it not a fair presumption, that the 2,275 men who constitute the difference of these numbers have been fraudulently rejected? We can readily understand that it is an easy matter to reduce one's height a half-inch, or an inch, when so great an interest is at stake as that of being rejected. On the other hand, the authorities are indulgent in cases where men of a more suitable size are substituted for small men. This conjecture becomes still more consistent, if we observe that 2,275 men are wanting to complete the two groups which are immediately above the limit 5 feet 2 inches.

The consideration of the probable error has this advantage, that it permits us to establish comparisons, even between series of observations which may not be homogeneous. If we were to submit to calculation the example in my thirteenth letter on the subject of the mean temperature of the month of July, we should find that the probable error is $3\cdot78°$; but we have also found the probable error in the measurement of the right ascension of the Polar Star to be $0^{s}\cdot8$. It is then an even bet, on the one hand, that an observation of the right ascension of the Polar Star made at Greenwich would not be $0^{s}\cdot8$ in error; and on the other, that the temperature of a July day at Brussels would be $66\cdot2°$, without the fear of differing more than $3\cdot78°$. It is well understood that if the circumstances were varied, (if, for instance, the locality were changed,) the conditions of the bet would be similarly altered, for the probable errors would no longer be the same.

When we have a series of observations expressed in a foreign measure, we are sometimes at a loss how to substitute another series, adapted to a scale of measurement with which we are better acquainted. Thus the circumferences of the chests of the Scotch soldiers were expressed in English inches; but a Frenchman would wish to know how the measurements would have been grouped, if we had proceeded by differences of—say two centimètres. A German would have preferred grouping them by differences of zollen.

These new arrangements can be easily made by means of the

scale of possibility. We may remark, in fact, that the numbers given furnish every means of constructing the curve of possibility. When we have this curve, it is easy to substitute, for a series of equidistant perpendiculars which divide its surface into bands of equal breadth, another series of perpendiculars also parallel one to another, but enclosing greater or smaller spaces. In using the table of possibility, we attain our end more easily by calculation.

There is a problem which frequently presents itself in applications of this kind,—it is as follows. A certain number of measurements have been collected, and we know the smallest and the greatest, which serve as limits to the series: it is required from these data, only, to group all the other observations. It would seem at first that these data are insufficient; it is however not so, for we have all the necessary elements for solution. Thus, let us suppose the tallest man measured in France to be 6 feet 11 inches, and the shortest only 3 feet 8 inches, and the number of men observed to be 1,000,000: it will be possible for me to say how these men are grouped with regard to height, whether I divide them into groups by differences of half an inch, or in any other manner. It is first evident that the mean size will be equally distant from each of the two extremes, and will consequently form their arithmetical mean. This mean height is 5 feet 4 inches nearly; and of 1,000,000 men measured, 500,000 would exceed, and 500,000 be less than this height. When distributed in ten groups of differences of 3·93 inches, our table of precision gives for the five groups which exceed the mean the following values placed at sixteen ranks apart:—

344,335	men of	63·7 inches	to	67·63 inches.
478,586	„	„	„	71·56
498,814	„	„	„	75·49
499,974	„	„	„	79·42
499,999	„	„	„	83·35

This table shows us that the men would be classed as follows in order of size,—

344,335 being 63·7 inches plus 3·93 inches.
134,251   ,,   ,,   ,,   7·86
20,228   ,,   ,,   ,,   11·79
1,160   ,,   ,,   ,,   15·72
25   ,,   ,,   ,,   19·65
1 being more than 83·35 inches.

The same distribution holds good for the five groups below the mean; so that we should find

344,335 men of 63·7 inches less 3·93 inches.
134,251   ,,   ,,   ,,   7·86
20,228   ,,   ,,   ,,   11·79
1,160   ,,   ,,   ,,   15·72
25   ,,   ,,   ,,   19·65
1 of less than 44·05 inches.

LETTER XXII.

ORDINARY AND EXTRAORDINARY EVENTS.—MONSTROSITIES.

Your Highness is perhaps already convinced of the utility of the consideration of limits, and of the mean,—you may have seen that all our ideas which aim at the appreciation of magnitudes generally repose on these three things. Often we proceed further; and, in addition to the two extreme limits, we unintentionally recognise intermediate limits also. Thus we use with advantage the *limits of probable error*, which are very narrow, since they include but a half of the entire number of observations: we may take them as including the *ordinary* cases. Beyond these limits are presented those cases which, without being extraordinary, give rise to the distinction of *great* or *small* in ordinary language. Here we find new limits, with which science has not yet occupied itself, and whose use is determined in the most arbitrary manner. Beyond these second limits are presented *extraordinary* cases, which are themselves limited by the impossible.

You may suppose that these limits should play a great part in our mode of viewing phenomena relative to man and to society. The different limits have probabilities which correspond with them: I am about to endeavour to define them, without deviating more than necessary from preconceived ideas.

I will first remark that we shall have the following distinctions, in the application to the height of man of the limits which I have established:—

Extraordinary limit in excess . .	*Giants.*
Ordinary limit in excess . .	*Large men.*
Limit of probable error in excess ⎱ Limit of probable error in defect ⎰	*The ordinary size.*
Ordinary limit in defect . .	*Small men.*
Extraordinary limit in defect .	*Dwarfs.*

I find then three kinds of limits: and in this I but conform to usage; only I seek to give more exact appreciations to things to which every one, according to his caprice, gives the most arbitrary values.

The limits of the probable error in excess or defect correspond to the probability $\frac{1}{2}$. We must not forget that the number of observations comprised within these limits is proportional to this probability, and forms the half of the total number of observations.

The ordinary limits may extend on either side of the mean, as far as the probability 0·4999999; so that of six millions we can only look upon two as extraordinary,—the one in excess, the other in defect. In France, for instance, we may regard as giants the three or four tallest men who are in the kingdom, and as dwarfs the three or four smallest, supposing them regularly formed.

The extraordinary limits, beyond which are found *monstrosities*, seem to me difficult to fix. If, in the scale of possibility which we have constructed, supposing a very great but determined number of chances, we take the limits somewhat distant from the mean, the corresponding probabilities become so small as to be inappreciable. When we suppose the number of observations infinite, we may carry the differences to equally infinite distances from the mean, and find the corresponding probabilities. This mathematical conception evidently cannot agree with that which is in nature.

It has appeared to me, that in the consideration of heights, we may admit with the greatest confidence those limits which extend beyond the greatest or smallest size ever observed in man. Thus, relying only on authentic documents, we know that Frederick the Great had a Swedish body-guard whose height was eight feet and a half in Swedish measure, or 8 feet 3 inches English. On the other hand, according to Birch, there has existed an individual, 37 years of age, whose height was only 16 inches. Bébé the noted dwarf (the King of Poland) was taller.

If we take 5 feet 4 inches as the average height, the giant of the King of Prussia exceeded it by 2 feet 11 inches, and the dwarf mentioned by Birch was 3 feet 11 inches below it. Although this last measure seems to me to be exaggerated, yet I will admit it as the inferior limit; and I will also admit as the superior limit the mean augmented by the same difference,—3 feet 11 inches, or 9 feet 3 inches.   The height of man will have then as its extreme limits —17 inches, and 9 feet 3 inches.

Buffon (from whom I have borrowed the foregoing measurements) thinks that we ought to fix, as the limits of actual nature, the size of the human body as being from 2 feet 6 inches to 8 feet in height; and that although this interval is considerable, and the difference (he adds) seems enormous, it is however greater in some kinds of animals.*

I admit yet wider limits than those of Buffon, and I found them on observations reported by the celebrated naturalist.   We shall then have the following limits for the human height:—

	FT.	IN.
Limit of the size of giants	9	3
Do. large men	6	8
Do. of ordinary size in excess	5	6
Mean size	5	4
Limit of ordinary size in defect	5	2
Do. of small men	4	0
Do. of dwarfs	1	5

The mean which I have here adopted relates to France.   Half the men of this nation, at the age of the conscription, are comprised between the limits 5 feet 6 inches and 5 feet 2 inches.

Of 10,000,000 men there is but one found above 6 feet 8 inches, and but one less than 4 feet.

An infinite number of men would be required to arrive at the two limits 9 feet 3 inches and 1 foot 5 inches, which I have taken as the greatest possible; and we may consider as monstrosities all falling without these limits.

P. S. Since I wrote the preceding, I have had occasion to direct my own observations to a miniature of the human type, who has for some time occupied the attention of the two worlds: I speak of Charles S. Stratton, who has been pompously surnamed Tom Thumb. I took advantage of the good fortune that was offered me, and have

* *Natural History of Man*, article DWARFS.

been able to measure the different dimensions of the body of this celebrated dwarf, whom the United States have offered to the admiration of Europe. I say "admiration," for I think the different classes of society have never shown a greater eagerness to see any other man. Kepler, Pascal, Newton, if they could reappear on the scene of this world, would be nothing by the side of the illustrious American General. It is true that genius, however great it may be, is not visible: it fares better with sizes,—particularly with a small size, which wounds nobody's self-conceit. There is matter for all the world to praise, without restriction.

However (I blush to avow it) Tom Thumb did not fulfil my expectations. I could not help yielding an instant to the general feeling, and admiring what I was about to call the great man; but an inflexible measurement repeated to me three times, with a rigorous obstinacy, that the General was 2 feet 3 inches high, and consequently far within my inferior limit. He is placed at a nearly equal distance from the two limits between which are included dwarfs of different sizes.

It will also be observed that these limits are fixed for adult dwarfs; and that the General Tom Thumb is but 14 years old, according to the avowal of those who exhibit him to the public. I would willingly take off one-half of his years, if I did not fear to affect his reputation. I should, moreover, declare that he is admirably well formed in his little dimensions: I have rarely seen more graceful hands, more pretty feet. Comparing the different measurements I was able to take, with those I had collected of children of different ages, I have found that Tom Thumb, at the time I measured him, had nearly the dimensions of an infant of 14 or 15 months; the arms, however, were notably shorter. The head, and all the extremities, which are generally much developed in dwarfs, in this case presented no extraordinary appearance. Only, the nose turned up, the sockets of the eyes strongly marked, a certain character of malice and penetration, gave the whole physiognomy a peculiarity which has been admirably rendered by Titian, in the beautiful picture in the Louvre, in which is seen the dwarf of Charles Quint. This picture in some measure portrays all the qualities of the dwarf; and if it is not a faithful portrait of Tom Thumb, it is because the type which it presents is of a more advanced age.

It is much to be regretted that exact measurements have not been kept of all. men remarkable for physical anomalies, particularly anomalies of size. It would be interesting to collect the measurements taken on different parts of the body, and to be enabled to discover what proportions are particularly affected by these anomalies. We generally confine ourselves to merely echoing the public voice, which is always disposed to exaggerate circumstances of this kind. It is more than time for science to occupy itself in the transmission to our descendants of more exact information of things which may merit their attention. Since the commencement of the practice of registering with precision those elements of the great phenomena of nature which are susceptible of calculation,—meteorology—astronomy—and the other physical sciences have of a truth singularly lost their character for marvels: imagination may have suffered, but reason is bettered. It is more than time that poetry also should mingle a little less with facts relative to man, and should permit us to see each thing without forcing our visual rays to pass through its deceiving prism.

THIRD PART.

ON THE STUDY OF CAUSES.

————

LETTER XXIII.

ON CAUSES, AND THEIR APPRECIATION.—CONSTANT CAUSES.—VARIABLE CAUSES.—
ACCIDENTAL CAUSES.

THE phenomena which organized beings present are so variable that they never perhaps manifest themselves under perfectly
identical circumstances.   We can easily imagine this, if we regard
the infinite number of causes which give them birth, and all the
degrees of intensity of which these causes are susceptible.

The flower which has just opened has only arrived at its state
of bloom and splendour by being favoured with a genial temperature, and by the moisture of the earth and the air.   Could it spring
up under the combined influence of these two powerful agents
of nature alone, in how many ways this influence may vary, even
during the course of a single week!   Since the beginning of the
world, the temperature and the humidity of the atmosphere have
perhaps not been twice in identically the same circumstances for
eight consecutive days.   What then shall we say, if we consider
that the blowing of flowers may depend on the action of the winds,
on their exposure, on the solar radiation, and upon a crowd of
other causes?

It is this extreme complication which renders the observation
and classification of phenomena so difficult.   If the causes which
operate on man to disturb his organization, and to surround his

frail existence with maladies, were not so numerous and variable as they are, medicine would long since have become a science. Must it be said, that we renounce all hope of one day seeing it reach this high rank?

The phenomena which inanimate bodies present are equally under the influence of numerous and variable causes: we have nevertheless been able to discover these causes, and to study their principal laws. Although all the resources of analytical mathematics are not sufficient to calculate rigorously the movements of the most simple machine .in the hands of the workman, or the course of the ball projected by the artillery-man, it is yet possible to arrive at results so close to those observed that we may neglect the difference, and make much use of the approximate results.

It appertains but to superior minds to perceive, in the midst of an infinity of causes which give rise to an event, those which are most influential, those which must be regarded, and those which may be neglected.

Without being ambitious of so high a mission, we may nevertheless seek to introduce into researches a greater degree of order; and we shall, perhaps, have secured the first step, if we can render to ourselves a good account of the nature of the causes which influence events.

These causes may be divided into three principal classes,—

Constant causes.
Variable causes.
Accidental causes.

Constant causes are those which act in a continuous manner, with the same intensity, and in the same direction.

Variable causes act in a continuous manner, with energies and tendencies which change either according to determined laws or without any apparent law. Among variable causes, it is above all important to distinguish such as are of a *periodic* character, as for instance the seasons.

Accidental causes only manifest themselves fortuitously, and act indifferently in any direction.

If we appreciate causes in a mathematical sense, the constant cause has in its favour a certain *definite* number of chances, a fixed probability.

The variable cause has in its favour a *variable* number of chances, and consequently a probability which may oscillate within greater or smaller limits.

The accidental cause, strictly speaking, has no chances in its favour, but it influences the order of the succession of events. Thus, in drawing balls from an urn containing equal numbers of white and black balls, which have their respective probabilities of being drawn, accidental causes do not introduce any new ball either of the same or of another colour, but they render the order of drawing more or less regular, and the variations from the calculated order greater or less, although in the long run their action is neutralized. In short, we can imagine that the balls may be mixed in an infinite number of ways.

Were constant causes only to exist without the admixture of accidental causes, they would from the first tend to the results which appertain to them in a ratio equal to their degree of energy; but as it is impossible to avoid accidental causes altogether, it is only after a number of experiments, which depend on the quantity and magnitude of accidental causes, that the results are in relation to the degrees of energy of the causes which produce them.

LETTER XXIV.

ON CAUSES IN GENERAL,—MORE PARTICULARLY ACCIDENTAL CAUSES, WHEN THE CHANCES ARE EQUAL.

THE distinction which I have established between the three kinds of causes, which in general influence all phenomena, will be better illustrated by an example. I will suppose the problem to be the measurement of a certain length,—say the height of a man; and in order to remove difficulties, I will at once admit that the measure of which we propose to make use is perfectly exact, the number of divisions being so limited that we can only appreciate the 250th part of an inch. I will even admit that the man whose height is sought stands quite motionless during the experiments, and has no tendency either to increase or diminish his stature. I will also grant that the person measuring uses all imaginable precautions to attain great precision; in a word, that there is no kind of cause, either *constant* or *variable*, which can affect the exactness of the results.

Notwithstanding all these concessions, we can conceive that the several measurements that shall be taken will not be identically the same: small differences between them will appear, influenced by *accidental* causes; and, without being unskilful, we might differ from the true measure which we wished to determine by the 50th part of an inch (for example) in excess or defect. Thus, whilst measuring, the rule may not lie evenly on the top of the head,—it will not always be supported with the same degree of force, nor always on the same point; the hair will become disarranged, and will form a variable thickness; the sight will not always be equally correct, nor the readings of the measure equally certain.

Those who are engaged in such measurements will doubtless grant willingly that, when the errors do not exceed the 50th part of an inch, they may be satisfied with the result obtained.

But if a thousand successive measurements had been taken, very few would be a 50th part of an inch in error. The greater part would differ from the true size, either in excess or defect, more than the 250th part of an inch, others the 500th part, others the 750th part, and so on.

In considering only the greatest and the smallest measurements, we should find between them a difference of as much as the 25th part of an inch; and we may conceive nine intermediate groups of measures, proceeding by differences of the 250th part of an inch.

These groups would not be composed at hazard, of numbers more or less great, although they arise from awkwardness or from defect in sight. Their formation is indicated à priori in the scale of possibility, relative to the drawing of ten balls from an urn, which should contain only black and white balls in equal quantities. Thus we may say beforehand how many observations will fall in the group which includes the smallest measurements,—how many will fall in the group immediately above, presenting a 250th part of an inch more,—how many observations will form the third group, of a 500th part of an inch more; and so on.

This singular result always astonishes persons unfamiliar with this kind of research. How, in fact, can it be believed that errors and inaccuracies are committed with the same regularity as a series of events whose order of succession is calculated in advance? There is in it something mysterious, which however ceases to surprise when we examine things more closely. The mode of action of accidental causes has not been studied with sufficient care; but, nevertheless, it presides in some measure over the explanation of all phenomena which depend on the calculus of probabilities. I have devoted to it a considerable portion of my former letters, and therefore think that I may without prejudice refrain from again dwelling upon it here.

So much for accidental causes. We see that in the long run a sort of equilibrium is established between errors in excess and errors in defect,—an equilibrium by which we arrive definitively at the true magnitude we wish to measure.

It will not be the same with a *constant* cause of error. This cause should prevail definitively; and not only will the final result be affected, but, under the influence of accidental causes, the different groups will be classed according to new laws, and will not have that symmetry which we have recognised when the chances were equal. Thus we have supposed that the person remained perfectly quiet during the experiments, and had not any tendency to lengthen or shorten his height. But, were the contrary the case, we should find ourselves exposed to fresh errors, in addition to accidental ones.

We may here make two hypotheses,—either that the person measured, without preserving an absolute immobility, has as much tendency to increase as to diminish his height; or that he has a greater tendency in one direction, whether it be to render himself taller or shorter.

In the first case, the person's motion ought to be considered as a cause of accidental error, whose effects will in the end be destroyed; in the second case, on the contrary, we shall definitively arrive at a measure either too great or too small. Were the measurement of the same person repeated day after day, for a sufficient number of times to destroy the effects of accidental causes, the constancy of the result we should obtain might lead us to think it exact. It may hence be seen with what circumspection we should proceed in the appreciation of magnitudes.

It would oftener occur that the mean height obtained would vary from day to day, because the person would not have the same tendency to increase his height. The cause of error would then be *variable*.

The cause of error may be variable in different ways. It may be so from the fact that the person measured, either from caprice or from other motives, has a tendency constantly to exaggerate his height; but not in a *regular* manner. It may again vary *periodically*, if the person be measured at different times of the day. It has indeed been supposed that man is taller in the morning, immediately after rising from his bed, than he is in the evening, after the fatigues of the day.   Thus a thousand measures taken in the morning would give a greater mean height than a thousand measures taken at night; and the height would vary by intermediate gradations at the intervening hours of the day.   Causes then may vary regularly and irregularly.   *Regular periodical* variations are very numerous in nature, and should occupy a special place in my letters.

LETTER XXV.

ON ACCIDENTAL CAUSES, WHEN THE CHANCES ARE UNEQUAL.

I WAS one day walking in the Gardens of the Trianon with Mr. George Edwards: our conversation turned upon the sciences of observation, and on the difficulties they presented in certain cases where the appreciation of means was involved. At this period, this *savant* was actively engaged on his dynamometrical experiments, and with the variations of strength which man experiences at different parts of the day, and which may be said to be consequent on his slightest actions. " Have you remarked," he said to me, "that in quantities which admit of a mean in their accidental variations, there is often a tendency to produce greater variations above the mean than below, although the chances of producing these variations seem to be absolutely equal?" I not only replied that I had made the same remark, but I recounted to him several examples which had struck me.

Nature is like man in this,—when it differs from its type, it is more generally in exaggeration than in diminution. This peculiarity continued to occupy us some time afterwards; and Mr. Edwards proposed to combine our observations and reflections on this interesting subject, to produce a common work. But soon after I was compelled to quit Paris, and our projects were lost sight of. I know not whether he continued to pursue the subject,—as for myself, I have often reverted to it; and I perceived that we had too much particularized the question, because all the examples with which we had been occupied carried the error in one direction.

It may indeed happen that, under certain circumstances, the tendency which we have observed may act in a different direction,

I

and that the variations below the mean may be generally greater than those above.

I should also add that this tendency to produce greater variations, either above or below the mean, when the chances of variation seem equal, is always small.

The curve of possibility then loses its form, and takes that indicated in the figure.

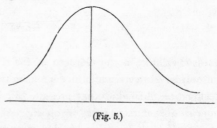

(Fig. 5.)

We find an example of this in the appreciation of the daily variation of the temperature during the winter months. The greatest variations fall above the mean.

The contrary takes place in barometrical observations. It has long been noticed that the falling of the mercury below the mean is in general greater than its elevation above it.

Like examples do not only present themselves in Physics,—they are also to be found in the political sciences. I have already had occasion to notice a very remarkable one concerning the price of grain. In the fluctuations which these prices experience, the differences above the mean may extend much further than those below this same quantity; and we may easily account for this circumstance.

Mortality is also a variable quantity, which in its extraordinary variations may exceed the mean much oftener than it rests below it. We do not remark that we meet with the inverse of pestilences or epidemics,—that is to say, that we meet with years in which the mortality diminishes considerably under the influence of certain causes.

Examples in which the mean does not fall at an equal distance from the extreme values, and where the curve of possibility loses its symmetry, are very frequent: they deserve to be studied the more, as this want of symmetry arises from causes more or less curious, whose influence we are able to appreciate. These causes are not of the nature of those which we have hitherto examined. Let us try to take account of them.

I will suppose that by use of compasses I wish to find the diameter of a spherical body, and that by different trials I find it has a diameter of 12 inches. Among all the measurements taken, there will be one greater than the rest, and one smaller. I have stated that these two variations from the mean should be equal, if the chances of error in excess and in defect were perfectly equal, and if the measurements were sufficiently numerous. I will here suppose them only to be one-tenth of an inch. Such should then be the greatest error I should have to fear.

Let us suppose that I employ the same compasses, the same mode of procedure, and that all the circumstances remain otherwise identically the same as to the exactness of the measurement. The result will be that, whatever may be the object measured, I shall obtain the same variations, and that in the long run they will re-occur.

We must however make an exception, if the globule to be measured is less than one-tenth of an inch in diameter: for, at the worst, I may indeed find among all the measures one diameter extremely small; but it would be absurd to suppose it below zero. In such circumstances the greatest inferior variation is no longer possible, and becomes necessarily less than the superior variation, which continues to be one-tenth of an inch.

It is not even necessary that the object to be measured should be less than the greatest inferior variation, to cause an inequality in the order of errors, and to destroy the regularity of the curve of possibility.

We therefore conclude that fortuitous causes, some of which tend to give greater, and some smaller values than the mean, are no longer really the same.

Our senses perhaps deceive us more in this respect than our instruments. We have a curious proof of this in the history of Astronomy, in regard to the valuing the apparent diameters of the stars. The first astronomers who occupied themselves with deter-

mining the angle under which the most brilliant stars which
radiate from the heavens were seen much exaggerated its size.
Kepler attributed to Sirius an apparent diameter of 240 seconds;
Tycho Brahe, more than 126 seconds; and Albategnius, before
them, made it 45 seconds. The diversity of these measures shows
sufficiently how little confidence they should inspire. The inven-
tion of lunettes shows that these measurements, even the smallest,
were much exaggerated. Gassendi gave Sirius but 10 seconds of
apparent diameter. Galileo, Hevelius, and J. D. Cassini reduced
this diameter to 5 or 6 seconds. Sir W. Herschel went even farther, and only estimated at small fractions of a second the apparent diameter of the two most brilliant stars in the heavens.
This celebrated astronomer even doubted whether the value for
Arcturus should not fall below the tenth part of a second. What
a prodigious difference in the estimate of stars of the first magnitude, to descend from 240 seconds to the 2,400th part of that
value!

Details relative to these different measurements have been presented, with as much interest as science, in a notice on the works
of Sir W. Herschel inserted by M. Arago in the *Annuaire du
Bureau des Longitudes* for 1842. It might be asked to which of
all these appreciations we should definitively assent? The perfec-
tion of the instruments, and the methods of observation, should
naturally incline us to modern appreciations; but we may add to
these motives for preference a species of direct demonstration,
deduced from the manner in which occultation of stars occur when
the moon passes before them. The moon has in the heavens a
sufficiently rapid motion to make it pass through an arc of half a
second of arc in a second of time. But if a star had an apparent diameter of half a second, it would require an interval of a
second to be entirely eclipsed by the moon. We should then see
it during this second *progressively* lose its light. But nothing of
the kind takes place; and all those who have observed occultations
of stars by the moon know that the disappearance is instantaneous,
and without previous loss of light. It is the same with the emer-
sion,—the star re-appears suddenly in all its splendour. This
simple observation, which can be so much the better verified as the

glasses employed are the more powerful, fully confirms the appreciations of modern astronomers.

We are not yet sufficiently acquainted with the extent of our faculties, nor with the limits of error to which they are subject, to be able to determine the chances we may have of being deceived in the different valuations which depend upon them.

The example which I have just quoted is curious in more than one respect. Thus, it might be asked why I do not take as the angular diameter of a star a mean between all the values given by the several astronomers? I reply, that to do so would be to place all the observations in the same rank, and to attribute to all the same value. If I have taken a different course in my preceding letters, it is because I have supposed that all the observations employed have merited an equal degree of confidence.

It is already easy for us to understand how it sometimes happens that, while considering errors in excess and in defect as being equally probable, we find results quite contrary to our hypothesis: our instruments and our senses deceive us without our knowledge, or we may have thought chances equal which were not really so. This is also exhibited in the very manner in which the results group themselves. So soon as the observations are influenced by causes which act more strongly in one direction than in another, the curve of possibility loses its regular form: the summit is displaced according as the causes operate to render the error greater in one direction than in another.

We may, for greater facility, group the causes of error into two classes,—the one tending to produce errors in defect, the other to produce errors in excess. I have examined at great length the case where causes have on either hand the same intensity: this is in fact the most interesting. The case of the two groups acting with unequal degrees of intensity requires, for its examination, that we should well understand the nature of the causes which influence the different events.

LETTER XXVI.

YOUR Highness will doubtless not have forgotten what I said on the admirable symmetry which would be manifest in the registrations of births, if boys and girls were born in equal numbers. Laws not less remarkable, although different, regulate the order of births, if boys are born in greater numbers than girls; and these general laws regulate the order of the occurrence of all events which depend on two kinds of unequal chances.

To reduce the question to its most simple terms, I will have recourse to my habitual oracle. I will interrogate an urn, which contains balls of two colours, but in unequal numbers. I will suppose, for example, 3 white balls to 2 black. I will further suppose that the balls are taken out singly, and that each one after the drawing has been entered is replaced in the urn.

After a thousand drawings, if any one had the curiosity to count the number of white balls and of black balls taken out by hazard, he would find two numbers which would differ but little from 600 and 400; and this ratio of 6 to 4 would be so much the more verified in proportion as the number of drawings is increased. Whatever might be the ratio, it would in the end be exhibited, were the drawings sufficiently prolonged.

In the same way also, although we do not know the secret of Nature, we may divine it with an ever-increasing approximation, by having the patience to interrogate her for a sufficiently long period. Thus, supposing that Nature produces male and female births in a determined and immutable ratio, we should become acquainted with the fact by sufficiently multiplying our observations. Not only would the births, taken one by one, present numerically this ratio, but by taking them two by two, three by three, &c., we might assign in anticipation the order of occurrence. But let us return to the example which I selected in the first instance.

If we take the balls two by two,—the first with the second, the third with the fourth, and so on,—we should form three sorts of binary groups; and 25 groups taken from the urn would generally be distributed as follows,—

> 9 groups of 2 white balls.
> 12   ,,   1 black and 1 white.
> 4   ,,   2 black balls.

If we take the balls three by three,—the first, second, and third, the fourth, fifth, and sixth, and so on,—we should form four kinds of ternary groups; and 125 groups taken from the urn would generally be thus distributed,—

> 27 groups of 3 white balls.
> 54   ,,   2 white balls and 1 black.
> 36   ,,   1 white ball and 2 black.
> 8   ,,   3 black balls.

If the balls were taken four by four, five kinds of groups would be formed; and 625 groups would generally be distributed thus,—

> 81 groups of 4 white balls.
> 216   ,,   3 white balls and 1 black.
> 216   ,,   2 white balls and 2 black.
> 96   ,,   1 white ball and 3 black.
> 16   ,,   4 black balls.

Without proceeding further, we perceive that our groups no longer present that symmetry which we observed when the chances of drawing the two colours were equal. We must not, however, conclude that there no longer exists any order in the results. The curve of possibility, without being symmetrical, preserves a very regular form: thus for 100,000 groups of 16 balls each, the distribution of the balls will be as is indicated in the fourth column of the table contained in the note upon this letter inserted at the end of the volume,—and the curve of possibility would have the following form:—

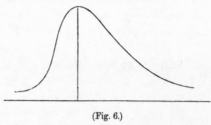

(Fig. 6.)

The group which has the greatest probability in its favour is the seventh: of 16 balls, it contains 10 white and 6 black. Theory, in fact, indicates that the greatest term is that which contains black and white balls in a ratio equal to that of their respective probabilities of being drawn. This ratio would here be 3 to 2, or 9 to 6.

We therefore see that the two limits are unequally distant from the *maximum* term, and consequently from the mean of the results. The mean here presents itself between the seventh and eighth term, but nearer to the seventh, which is the consequence of what I have above stated. The differences between the mean and the two limits are inversely proportional to the chances in their favour.

The mean will approach more or less to the one or other limit, according as the chances for or against the expected event are more or less unequal. The table which accompanies this letter may make this more sensible. (*See* notes.) Ten vertical columns successively indicate all the probabilities of drawing 16 balls at a time, in an urn containing an infinite number of white and black balls, allotted so that there may be, out of 100 balls, 50 white or 55, or 60, and so on.

The first horizontal column expresses the probability of not drawing any black balls, and consequently of drawing only white balls. The second column gives the probability of drawing only 1 black ball and 15 white; the third, of 2 black and 14 white; and so on to the last, which expresses the probability of taking 16 black balls at a time. Probabilities superior to ·00001 have alone been inserted.

It will be remarked that the *maximum* probability which occupies the ninth place, where of 100 balls the urn is supposed to contain 50 white, successively rises and takes the first place, where out of 100 balls the urn contains 95 white and 5 black.

The lines of the following diagram will render this result more sensible. Each one represents the numbers of one of the vertical columns; and as we can make our table so as to express all imaginable combinations, it will be easy to conceive how the series which I am compelled to omit would proceed.

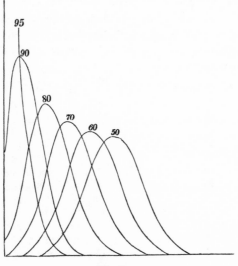

(Fig. 7.

If I have been sufficiently clear in the preceding remarks, the following conclusion may be drawn from them. A very great number of balls having been successively drawn from an urn which contains as many white balls as black,—and these balls having been taken 16 by 16, for example,—the groups would be symmetrically distributed around the greatest group when I separated them in order of colours. If the urn contained more black balls than white, there would no longer be any symmetry: we should still find a maximum group, but the other groups would be unequally distributed on either side of it. This inequality would be so much the greater as the two kinds of chances offered greater numerical differences.

I will take an example from among those I have mentioned in my previous letters. The temperature at Brussels varies sensibly in the space of twenty-four hours: these variations, however, are smaller in winter than in summer. I have already said that in December and January the daily variation of the temperature is from four to five degrees centigrade (or seven to nine degrees Fahrenheit), whilst in June it is double.

But let us take the variation of January: it has not always been from four to five degrees centigrade,—it has sometimes been greater and sometimes less; and among 309 days of observations made from 1833 to 1842, these variations have been distributed in the following manner :—

			CENTIGRADE.			FAHRENHEIT.		
8	daily variations from		1	to	2	1·8	to	3·6
31	„	„	2	„	3	3·6	„	5·4
61	„	„	3	„	4	5·4	„	7·2
68	„	„	4	„	5	7·2	„	9·0
50	„	„	5	„	6	9·0	„	10·8
32	„	„	6	„	7	10·8	„	12·6
22	„	„	7	„	8	12·6	„	14·4
20	„	„	8	„	9	14·4	„	16·2
8	„	„	9	„	10	16·2	„	18·0
4	„	„	10	„	11	18·0	„	19·8
3	„	„	11	„	.12	19·8	„	21·6
1	„	„	12	„	13	21·6	„	23·4
1	„	„	13	„	14	23·4	„	25·2

Although the numbers in the first vertical column are not symmetrically distributed on the two sides of the greatest number, it is however easy to perceive that they decrease in a very regular manner. If we had constructed the line which would represent them, it would have resembled considerably in form that represented by the numbers in the seventh column of our general table. This seventh column expresses the chances of drawing from an urn which contains 80 white balls and 20 black in every 100. The analogy would be still greater, for the numbers given above, if the urn contained 81 white balls and 19 black. What must we conclude from this relation ?

The conclusions are the following :—

1st. There exists a daily variation of temperature of 4°·7 centigrade (8°·46 Fahrenheit): this is given by the mean of all the observations.

2nd. This variation is subject to the influence of unequal causes.

3rd. The causes which tend to lower the daily variation to its *minimum* have more chances in their favour than those which tend to raise it to its *maximum;* and the chances are in the ratio of 81 to 19, or nearly 4 to 1.

4th. The distances of the mean from the two extremes are regulated by the same ratio of 4 to 1.

This last result does not appear quite to accord with that given by experience. For *above* the greatest term 68 there are nine, and *below* three. These numbers are in the ratio of 3 to 1, and not 4 to 1. But a greater number of observations are required to establish a complete accordance between theory and experience.

In winter the causes which tend to reduce the daily variation to its inferior limit have a much greater probability than those which act in an opposite direction. In the month of June, on the contrary, these latter causes have rather more chances than the former, and it passes gradually from one state to the other. This will be better understood by a second table, in which are given all the daily variations observed each month, and the corresponding proportionals, in order to facilitate comparisons. (*Vide* notes.)

The two inferior horizontal columns show the relative influence of the two kinds of accidental causes which tend to make the normal value oscillate between its two limits.

Among accidental causes, we may then consider the one as tending to bring the variable to its inferior limit, and the other as tending to elevate it to its superior limit. When the causes are equal to one another, the curve of possibility is symmetrical: when the one has more chances in its favour than the other, it draws towards itself the summit of the curve; and the approach is so much the greater as the number of chances is more considerable. Sometimes the chances are not subject to any appreciable law, and the curve of possibility may assume the most capricious forms. What I shall presently have occasion to say, on the nature of causes, renders it unnecessary for me to enter into more ample details on this subject, for the present at least.

The accidental causes which affect the price of grain may be ranged in two classes,—the one tending to its elevation, and the other to its depression.

Were the former causes to act simultaneously, without interference of the second, grain would attain its *maximum* price. On the opposite hypothesis grain would fall to its *minimum* value; and the two limits would be thus attained.

But it generally happens that the causes which tend to elevate the price, and those which tend to depress it, partly neutralize one another; and the most probable price grain will reach will be that which is distant from the extreme limits in the inverse ratio of the energy of the two classes of causes. Thus, from the price of grain, in its variations from its normal state, accidentally rising more than it falls, we may conclude that the accidental causes which tend to bring it down to its inferior limit have a greater probability,—or in other words, greater energy than the contrary causes.

It is the same with mortality: the accidental causes which tend to bring it down to its inferior limit have much more probability than the contrary causes, either because they are more numerous, or because they are more energetic.

LETTER XXVII.

ON CONSTANT CAUSES.

IF I have devoted much space to the study of accidental causes, it is because this study merits much of our attention: even beyond the boundaries of science, it presents something mysterious, well calculated to excite curiosity. Is it not indeed remarkable, to see in a long series of experiments all the small causes of fortuitous errors compensate and neutralise one another? And the compensations are established in so regular a manner that the different results obtained, in the place of the true one sought, are arranged on the two sides of that one in an order which may be previously assigned. This order remains the same, unalterable: the limits only which embrace the results are more or less confined, according to the skill of the observer, and the more or less precise means he employs.

It is then possible to destroy fortuitous errors, and to have a measurement of the accuracy of the observer, as well as precision in the final result.

This result enables us to appreciate *constant* or *variable* causes. Theory in effect teaches us that, by sufficiently multiplying experiments, the opposite result occurs with respect to these causes to that which we remark for accidental causes,—that is to say, that their effects are more defined in proportion as the effects of these latter causes are effaced, and that definitively the event is modified in proportion to the energy of each cause.

These principles being of the greatest importance, I will present some developments to render them better understood.

Suppose it were required to determine in a *general* manner what is the probability that a child about to be born will be of the male sex. It is evident that, to reduce this appreciation to the easily calculable case, where all the chances are known, we must learn in what ratio male or female births occur. However, Nature has not transmitted to us its secret in this respect: she only permits us to seek to penetrate it by observation. It is to observation then that we must have recourse. By using it, with all the care and with all the precautions which science requires, we get the necessary data for the solution of the problem, and we find that for 100 female births there are generally 105 males.*

The question is reduced, then, to the same terms as if it related to the probability of drawing a white ball from an urn containing 105 white and 100 black,—or more strictly speaking, containing an infinite number of white and black balls distributed in the ratio of 105 to 100. This probability, as we already know, is $\frac{105}{205}$.

To compensate for Nature's silence, I have stated that we must recur to observation. But what is the course to pursue under such circumstances? In examining a question like that now before us, I have remarked, at the close of my first letter, that the observations collected by one individual (even by a professional man who carefully registers the results of his practice) would not be sufficient. From the two or three thousand cases which such practice may embrace, all that can be recognised is that the male births have differed very little from the female births. A second observer, quite as conscientious as the first, may have been enabled to deduce from his particular practice a result of similar character; except that perhaps he will have found a few male births in excess, whilst the opposite was the case with his fellow-practitioner.

The result would be that neither one nor the other has collected a sufficient number of observations whence to deduce a result representing the true course of nature, or the ratio between male and female births which generally obtains. Accidental causes were not properly eliminated, and their effects tended to disguise those of constant causes.

It is therefore necessary to observe on a larger scale. Well, let us seek to discover whether the results deduced annually, from the practices of all the observers in a great town, will suffice for

* *Physique Sociale*, vol. i., p. 43.

our purpose.   Our conclusions will indeed be so much the more precise, that we have in such case less to fear from the negligence of observers, since we take our data from official registers in which entries are made in virtue of a law.

The table of births in Brussels, from 1825 to 1842 inclusive, at once shows us that, during 18 years, the number of male births has only twice been inferior to the number of female births, and that once the two numbers were equal.*   Again, in the two first mentioned cases, the excess was scarcely sensible, the numbers being in fact,—

   In 1832 . . 1,851 males and 1,854 females.
   In 1839 . . 2,247 ,, 2,272 ,,

There was then a marked predominance in favour of male births. We might even calculate, from what has been stated in my previous letters, what would be the probability that the year 1843 also presented an excess of male births, and also that some cause exists which facilitates this inequality.

If we had followed this kind of observation from 1825, making our result more perfect as the observations accumulated, things would have been presented in another point of view.   Thus, for the year 1825, there were 2,004 male births and 1,759 female, which gives a ratio of 114 to 100.

This very large ratio would have been much modified by the results of 1826, which gave 101.   Which must we credit?   Good sense, according with science, counsels us to make fresh observations, and to adopt provisionally the mean result 107, which we should have obtained by combining the observations.

The third year would have again modified this ratio, by diminishing it.   We should have obtained 106 as the result of the three first years of observation.

Continuing to operate in the same manner, we should at the end of the tenth year have had as the ratio sought 104, and after eighteen years 105.   This last ratio, compared with those obtained each year separately, would have differed more or less : the greatest variations in excess and defect would have been exhibited by the years 1825 and 1839, when the ratios were 114 and 99.

  * *Annuaire de l'Observatoire Royal de Bruxelles* for 1844.

The precision of the general result 105, compared with that of any year, would have been in the ratio of the square roots of 18 and 1, from what theory has already shown us, or as 4·24 is to 1.

Although theory indicates, as a general position, that the precision of results augments with the number of observations, it may so happen that we accidentally fall, at the very beginning, on results more precise than we should obtain by adding new observations to those already made. Thus the separate results of the years 1829, 1831, 1833, 1834, &c., have all given the ratio 105, and approach nearer to the definitive result than the ratio calculated on the sum of the observations for the ten previous years. But this precision is only fortuitous, and there is no reason to trust in it with a sufficient probability.

If we continued the observations indefinitely, we should obtain a ratio which would approach nearer and nearer to the true ratio, of which we are ignorant, and of which nature makes a mystery.

The Theory of Probabilities gives us the means of calculating what is the probability that the ratio found does not differ from the ratio sought beyond a certain assigned limit. Thus, although we do not know what is actually the ratio of the number of male births to the number of female births, we may yet say that a certain number may bet against 1 that the ratio 105 to 100 does not differ from the true ratio by more than unity, for example. This I have already shown, when treating on the tables of possibility and precision.

There exists then a constant cause which tends to make male births predominate over female births: we may estimate that this cause is, to that which determines female births, as 105 is to 100 as to their respective intensity.

We here admit, as a general position, that *effects are proportionate to the causes which produce them.* This fertile principle serves (so to speak) as the foundation of all the sciences of observation.

Since it is so advantageous to make our observations upon large numbers, it is evident we should have done better had we taken, as the basis of our calculations, a number of births produced in a long series of years,—as the number of births produced in all Brabant, or (better still) in the Kingdom of Belgium. The following table gives the result of such a calculation; but, not to extend it too far, I only give the numbers which refer to five years.*

NUMBER OF MALE BIRTHS TO 100 FEMALE BIRTHS.			
Years.	Brussels.	Brabant.	Belgium.
1832	100	106·2	106·7
1833	105	103·7	108·0
1834	105	105·3	107·2
1835	109	105·5	105·6
1836	112	106·1	106·3
MEAN . .	106·2	105·3	106·7

The greatest variation from the mean was 6·2 for Brussels, 1·6 for Brabant, and 1·3 for the Kingdom of Belgium. The limits of variation are then narrowed in proportion as I operate on greater numbers. One very remarkable circumstance here presents itself,— it is, that the general ratio for these five years differs very sensibly, according as we take it for Brussels, for the province, or for the kingdom.

We should first ask ourselves, if this difference should be attributed to the effect of accidental causes, or to that of constant or variable causes. We should be inclined to the former hypothesis, if we consider that the single year 1836 throws a species of perturbation into the numbers at Brussels; in fact, by only taking the numbers of the four preceding years, we should have not quite 105 as the ratio, instead of 106·2. However, it is well to examine things closer, and to seek if some causes do not exist which make the ratio for Brussels differ from that of the kingdom, or even from that of Brabant only.

* Vide Les *Annuaires de l'Observatoire Royal de Bruxelles.*

K

These causes really exist; for, if we compare the births of the two sexes in towns and country places, we find that there are generally more boys born in the country. We should then have committed a grave error, in confounding two things which ought to be kept distinct.

This first example will render us more circumspect. We are already aware that we ought not to obtain the same ratio for different localities. The ratio varies not only with localities,—it is also influenced by many other circumstances. Hence it is only with great caution that we can enumerate the influencing causes, and appreciate their effects.

But the complete enumeration of the influencing causes is nearly impossible in the greater part of social phenomena, because these causes are not only very numerous, but sometimes they are so indirect, and at the same time so feeble as to elude all investigations. The property of an observing genius is to know how to seize upon the most influential causes,—those which sensibly modify phenomena,—those, especially, which act in a continued or periodical manner; and to abandon the others as not necessary to be regarded, and as ranging among the accidental causes, whose results are inappreciable when the trials have been sufficiently repeated.

This is the place to make a curious remark concerning errors almost inevitable, and of which perhaps sufficient notice is not taken. It concerns the observer himself, and the manner in which his observations are collected. I suppose, for example, that negligence causes the omission of registrations of births, without the omissions being deliberately made in one kind of births rather than the other; they will cause certain inaccuracies, sometimes in excess and sometimes in defect, which in the long run will destroy one another. For, according to our hypothesis, the chance of one error in excess is the same as that of one error in defect. This cause of error may then be ranged among the accidental causes.

If, on the contrary, the registrations were made with a certain propensity to omit the male births, (for instance, either on the part of the authorities, or on the part of the parents, to withdraw their children at a future period from the military service,) this propensity, acting always in the same direction, would in the end be manifest, and should be ranked among the constant causes,—or rather among the variable causes, since it would change with the chances of war or of danger.

It is these constant and variable causes which it is so important to recognise in social phenomena. In most instances, it is almost impossible to assign them *à priori:* their effects are only perceived in the numbers, and then it is very difficult to determine their true nature. Thus we see that there are generally more boys born in the country than in town; but what is the real reason?

The causes of the variation in the ratio of the two sexes are very numerous. I have examined successively the influence of the climate, of sojourning in towns and in the country, of the legitimacy of the births, of the age of the parents, &c. Among these different causes, it would appear that the age of parents is most influential; but we have as yet very few exact observations on this subject.

When many equally influential causes are thus presented, and it is wished to analyze the effects of each, the art consists in collecting all the good observations, and grouping them in such a manner that all the causes, except those whose influence we wish to appreciate, may be considered as having acted equally on the numbers of each group. If this separation has been made with sufficient care, and if we still notice differences between the results, we may attribute them to accidental causes, or even to the cause whose influence we seek to determine.

It is not always easy in such a case to avoid error, and not to attribute to the cause with which we are occupied the results of many causes which have acted simultaneously in the same direction. Appreciations, apparently the most simple, often give rise to the greatest errors.

LETTER XXVIII.

THE habit of observing has not with me blunted the sentiment of admiration which I have ever felt at the sight of the heavens. The magnificence and the imposing regularity of this spectacle contrast marvellously, in the calm of night, with the rapid and tumultuous succession of objects with which we have been occupied during the day,—one feels, so to speak, transported into another world. The silence of an observatory, the monotonous and regular beat of the pendulum, and the still more regular progress of the stars, add much to this illusion. We then better understand the weakness of man, and the power of the Supreme: we are struck with the inflexible constancy of the laws which regulate the march of worlds, and which preside over the succession of human generations.

Each second which I count is cut off from my short existence. I can already foresee the time when another observer will seat himself in the place I occupy, and will in his turn fix his attention on the star which I am observing. All around me will have changed, except these brilliant spheres—which, in the midst of so many vicissitudes, will preserve their unalterable march. But what do I say? Even these spheres may, perhaps, have an existence limited in time as their dimensions are limited in space; and I judge them eternal only by comparing them to the narrow bounds between which my own existence is placed.

The stars may pass through space with a prodigious swiftness, without appearing to my eyes to move. To judge of the great laws of nature, we must know how to place ourselves at convenient distances. Some of the distances do not seem to be entirely beyond our reach. Thus, there is in the constellation Cygnus a little star which annually changes its place by an extremely small apparent quantity, although in reality it proceeds with the greatest velocity. This star is probably one of the nearest stars to us; and if we cannot discover the same movement in the others, it is because their distance from us is more considerable.

Further, these celestial bodies in travelling through space have probably, as our sun, a rotary motion upon an axis, and a retinue of planets around them. This, at least, we are induced to believe with respect to *variable* stars, which have periodical returns of greater and lesser light.

Some stars have been observed to shine for a certain time, and then to disappear; whilst *nebulous* stars exhibit a curious spectacle of suns in process of formation, in a more or less advanced stage of development. If we have not been able to watch the whole process, it is because time has failed for our observations. What in fact are a few centuries in the consideration of the existence of worlds?

In the presence of these great phenomena, it may be asked whether there be any constant causes, and whether all things in the universe are not subject to fluctuations?

Compared with the innumerable suns which adorn the vault of heaven, our planet is but a very secondary body,—a grain of dust lost in immensity; and yet centuries have been required to bring it to the state in which we now see it. When we study it, we recognise the different successive phases through which it has passed, and obtain the conviction that on its surface there is nothing stable.

Nature has herself taken care to register, with an admirable fidelity, the phenomena which have succeeded one another since the origin of our earth; and she has preserved to us the types of the early species of plants and animals.

We see an entirely special organization in these odd products of a primary world, destined to live by turns in the midst of waters, in miry lands, or in a thick atmosphere. Some, with the gait of reptiles, are endowed with the organs necessary to swim; others spread out large wings from between the scales with which they are armed. Imagination alone could dream of forms so extraordinary, had not Geology found them laid up for ever in stone, wherever it hollows out the ground to ask for the annals of the primary world. Vegetation even is in accordance with these monstrous beings, whose wreck only now remains.

In producing these metamorphoses, Nature is far from wishing to conceal their secret. The book where she has inscribed them is constantly open before us. She has traced even in the tree which is developing itself the records of the meteorological variations it has undergone; and man's curiosity finds in the concentric layers of the wood the traces of the more or less severe winters which have preceded.

The Supreme has then not only spread life and movement throughout, and willed that its impress should be preserved, but has done more; for he has permitted man to associate in some degree with his work, and to modify it. We find, in fact, nearly everywhere on the surface of the globe the works of nature altered by the hand of man. I have elsewhere named those causes *natural* which influence phenomena, and which act beyond the sphere of our activity; and those *disturbing* causes which we ourselves develop, and which tend in general to alter the progress of nature.

These last causes are essentially variable: they bear in some measure the seal of human weakness,—they only exercise their power temporarily, and within very narrow limits. Natural causes, on the contrary, even in their variations, have a character of greatness and generality which demands our utmost admiration. We may oppose them, or modify them in their manifestations, but without ever altering their nature: they always in the end resume their original character.

To discover *variable* causes, the most simple mode is to divide into groups or series the objects supposed to be under their influence. When these groups are formed in the same manner, and are in all respects comparable, they will be successively equal one to another, if the causes which have given them birth be constant. On the contrary, they will be unequal if the causes be variable.

To take an example of this, I will inquire whether, according to the experience of late years, the mortality in Belgium has been subject to constant or variable causes. For this purpose, I take the returns of the mortality for the nine years from 1831 to 1839 inclusive, and will divide them into three groups, or three triennial periods, so as better to eliminate accidental causes. I find as follows,—

From 1831 to 1833 . . 1 death to 37·9 inhabitants.
„   1834 to 1836 . . 1   „   39·7   „
„   1837 to 1839 . . 1   „   37·9   „

The mortality for the first and third period was the same; but it was less during the second, which comprises the years 1834, 1835, and 1836. If we seek a cause for this inequality, we shall probably find it in the price of provisions. To assure ourselves of this, let us place by the side of the rate of mortality the prices of wheat and rye; bearing in mind the fact that the influence of scarcity or abundance is not manifested immediately, but generally in the returns of the following year. Consequently, in grouping the results into three-years periods, I shall take the prices from 1830 to 1838.

| TRIENNIAL PERIOD. | MORTALITY. | PRICE OF | |
		WHEAT.	RYE.
		s. *d.*	*s.* *d.*
1830 to 1832	37·9	17 0	11 8
1833 to 1835	39·7	11 7	7 6
1836 to 1838	37·9	14 2	9 2

We see that during the years 1833-4-5, the prices of wheat and rye were less than during the two other triennial periods. It would therefore appear that the actual cause of the diminution of the mortality during this same period was the reduction in the price of grain.

In researches of this kind it is well to proceed with the greatest circumspection, because we may easily displace causes, and attribute to one circumstance that which is due to another. But we

know that one of the causes which has the greatest influence on
mortality is found in the greater or less abundance of provisions,
and in the facility the people have of procuring them. The scarcity
of 1816 offers one of the most striking examples. This calamitous
year left profound traces in the lists of mortality of the following
year, as well as in the registers of marriages and births.

The influence of variable causes in general gives rise to many
difficulties in statistical researches. Instances are more particularly
found in researches in reference to population.   Thus the movement of a stationary population is often enough compared with the
movement of a population increasing by an excess of births over
deaths.   However, it is a comparison of heterogeneous elements.
All other things being equal, this latter population should have a
greater mortality, for there are more children it.

It is also evident, that in the countries where the population of
the towns is augmented at the expense of that of the provinces, the
mortality should be increasing, since the portion of the population
which is transported from the country into the towns is exposed
to greater dangers.   It would then be illogical to compare a like
population in its successive states, since it does not rest homogeneous from one year to the next.   From the deaths being more
numerous generally in towns than in the rural districts, it results
that two populations inhabiting different countries are rarely comparable to one another.

I have stated that causes may vary in different manners; and
in fact some vary without obeying any apparent law—eluding all
foresight : of this number are years of abundance and scarcity, and
all calamitous years in general.   Others undergo variations of
which we know the laws: they may be ranged in two principal
categories,—1st, periodical causes, as the seasons; 2nd, causes varying from a mean,—that is to say, those which only experience
accidental changes.   The examination of these two kinds of causes
merits particular attention.   I shall occupy myself with them successively in my following letters.

LETTER XXIX.

ON PERIODICAL VARIABLE CAUSES.

20th December, 1842.

AMONG the variable facts which we observe on the surface of our globe, the most remarkable are certainly those which obey the laws of periodicity. The phenomena, for instance, which fall under the influence of time are greatly modified by the diurnal and annual periods. These phenomena to this day have been considered separately, and have found places in the different branches of science according to the caprice of those who have observed them. It may, however, be easily conceived that this species of allotment must arrest the progress of science, and prevent us from discovering the general connections of periodical phenomena one with another. Thus, statistical science has examined with a scrupulous attention the influence of the seasons on mortality, on crime, on mental alienation, on suicides, on commerce, &c.: it has borrowed from Meteorology the indications of the temperature and variations of the atmosphere, and required from Medicine the results of its observations on the nature and severity of maladies in their relation to the different months of the year. It receives from the natural sciences, and from agriculture, information upon a host of interesting facts; but these facts are almost always collected and classed in separate tables,—no one, so far as I know, has thought of observing them simultaneously.

The idea of filling up this gap in science has long impressed me with the necessity of establishing as complete an enumeration as possible of periodical phenomena. I have thought it useful to submit it to the learned, with a view of better eliciting the importance of a subject whose object is not only to investigate general laws as yet little known, but also to present better means of appreciating climates, and of comparing them one with the other.

I have felt much flattered by seeing the proposition of such a system of observations favourably received by a great number of learned societies, and by some of the most illustrious men of the age. I shall probably soon have an opportunity of entertaining your Highness with the first successes obtained in this scientific crusade, which I preach under the patronage of our Academy,—at least for that part which relates to the natural sciences and the physics of the globe. I will not however delay submitting to your notice the part of the Report in which I have just published an account of the results we propose to obtain. I copy it verbatim.

"If life produces among individuals a series of phenomena which are modified every instant, and diversify themselves one with another; seasons and days, by their succession, do not exercise a less remarkable influence in simultaneously modifying not only the entire globe, but also all the living beings with which it is covered.

"It seems that *natural periodical phenomena* constitute a common life beyond the individual life. Their importance has often occupied observers; but they have generally neglected to study them as a whole, and to seek for the laws of dependence and co-relation which exist between them. The Academy has not feared to undertake this difficult study : we shall soon be able to judge if it was over-confident of its own powers, and of the esteem which it has obtained abroad. For some time since its researches have been directed to the study of the great atmospheric phenomena : without as yet forming the vast plan which was afterwards to occupy its attention, it was felt that, before all, the medium in which all living beings were plunged should be studied. To meteorological phenomena the physical phenomena of the globe naturally connected themselves. An active correspondence, which extended beyond the limits of Europe, furnishes the careful registration of the most remarkable events, such as aurora borealis, earthquakes, magnetic perturbations, hurricanes, waterspouts, &c.; and thus enables it to judge of the limits within which they are included, either in respect to time or space. It could thus recognise the relations, greater or smaller, which exist between them, and better appreciate the causes which give them birth. But no study engaged more serious attention than the progress of atmospheric waves.

" To the study of the periodical phenomena of the atmosphere was connected that of the periodical phenomena relating to plants, animals, and men. There is no one so great a stranger to the natural sciences as not to be struck with the magnificence with which Nature varies the physiognomy of our globe, according to the seasons, and even according to every instant of the day. This succession of phenomena, when attentively observed, is accomplished in the most regular manner and with the most striking harmony. If the eye could grasp the whole at once, it would see, after winter, vegetation develop itself progressively in our hemisphere from South to North, and unrol (so to speak) its verdant waves as day by day it extended its limits. But these limits, what are they? What hand bold enough to trace them on the globe? Besides, are they each year the same? Do they not vary according to the nature of plants? And when in their turn the flowers develop themselves, how are propagated these new waves of a sea embalmed and diapered with a thousand colours? What modifications do they undergo in their progress? In what order are the fruits produced? And the different animals which mix in this brilliant retinue, do they wait for a natural sign to show themselves? The birds especially, do they constantly follow the same route in visiting our climates? and is their existence dependent on the return of the same phenomena? How many different questions at once present themselves! and how worthy they are to stimulate the imagination of the naturalist, and to occupy his meditations!"*

The moral and political sciences, for their part, present to our curiosity problems no less interesting. I not only speak of all

* *Rapport sur les Travaux de l'Académie Royale des Sciences et Belles-Lettres de Bruxelles* in 1844. Read at the annual public meeting of the 15th December of the same year.

the fluctuations which the movement of the population undergoes under the influence of the different months of the year—of the alterations observed in the physical condition of man, even at the different instants of the day,—but of the manner in which morality and intelligence are affected. Who does not know that the severities of winter, by multiplying wants, cause in society a greater number of crimes against property, while they deaden the passions, which again wake with more ardour and danger at the return of the spring, and during the heats of summer? It is then that we see acts of violence break out—that revolt is organized and spread with more rapidity—that our intelligence, too ready to exalt itself, overleaps the last limits of reason. Singular condition of man and societies, that virtues and vices—that disorders of heart and mind —that public commotions—are influenced more or less by the distance of the sun from the equator, by the greater or less elevation of that luminary above our horizon!

These rapid indications will, I hope, suffice to justify the value I attach to periodical phenomena. In other respects this study presents numerous difficulties, because the cause whose influence we wish to determine is often intermingled with others which complicate the facts observed.

When the existence of a simple periodical cause is suspected, it is easy to study it, by comparing the different parts of the supposed period with one another. Thus, if we wish to know whether the mortality is influenced by the period of the year, we must compare the results of the different months of the year. We shall find that the mortality is subject, at an interval of about six months, to a *maximum* and a *minimum*. In our climate the *maximum* occurs in January, and the *minimum* in July: between these two limits the other numbers increase and decrease with regularity.

The influence of periodical causes may be easily eliminated by carrying the means which we compare throughout the whole period. Thus we had no difficulty in comparing the mortality of one year to that of different years, although the mortality varied considerably in the course of each. It is enough if the fluctuations are generally the same, which is in fact the case.

If we had no means of comparing the numbers belonging to

two complete periods, we could only establish comparisons between those numbers which belong to the same parts of the period. Thus, if for a particular country we were not in possession of the temperatures for two different years, we might nevertheless make a comparison between the temperatures of two corresponding months; or if we cannot observe the temperature at different hours of the day, we may at least always observe it at the same hour: and the results we shall obtain will allow of interesting conclusions on the influence of the seasons.

The periodical phenomena which occur within the limits of a day, or of a year, are not the only ones which should occupy our attention: there are others, such as the tides, which are more particularly influenced by lunar revolutions: there are others also, independent of astronomical causes. I will only quote one, because it is nearly unknown, and perhaps some students of Physics may be induced to consider it with more care than I have been able to do with a friend, to whom the science of Optics is indebted for the most ingenious researches. M. Plateau and I had proposed to revert to it, and to study it with especial care; but I have reason to fear that we can no longer realize this project.

The phenomenon consists of this. We know that when we have fixed our eyes for some time on a brilliant colour, and then turn them rapidly to a white ground, we see the colour which is *complementary* to that seen at first. Thus, to the impression of a *red* colour we may have observed succeeds a *green*. This colour fades, disappears, and returns to disappear again,—and so on, until it is completely lost; so that the *complementary* colour is not seen continuously, but intermittently. To make these experiments, one of us looked steadily, for a fixed number of seconds, at a piece of orange-coloured paper placed on a black ground, in a well-lighted place,—then turned our eyes immediately to a white wall. We then indicated, with the greatest possible precision, the instants when the accidental impression attained its successive *maxima* of intensity; whilst the other observer, with a watch with a second's hand, noted the times. The effect produced during these trials was confined to the disappearance and re-appearance of the accidental impression, without the return of the primitive impression. It was this interval of time which separated the successive re-appearances of the accidental impression that it was sought to determine. First, were the periods equally long? and then, were they equally numerous?

We found that the number of re-appearances was so much the greater, as the time during which the primitive object had been viewed was longer.

We also remarked that, although the time that we had viewed the object exercised an influence on the number of re-appearances, it did not appear to have any over the duration of each.

On comparing our results together, we found that the re-appearances had sensibly the same duration for the eyes of each of us. Only after having viewed the object for a long time, and when the periodical repetitions became frequent, some secondary repetitions which escaped one of the observers was seen by the other.*

These experiments, I repeat, were not sufficiently numerous for the conclusions to be adopted with all necessary confidence; but those who may be willing again to undertake them will find in them a curious example for appreciating the influences exerted on the sight of each observer, by the time he fixed his eyes on the colour submitted to trial, by the nature of such colour, &c.

The beatings of the pulse, breathing, certain practices of man, and many of his maladies, should also be referred to periodical phenomena; but I fear becoming wearisome, if my subject is too much extended.

I cannot, however, abstain from presenting some general remarks, which will not perhaps be without interest. I will confine myself to their enunciation.

1st. If all the causes which act in the universe were constant and uniform, our world would remain in an invariable state, life would be everywhere extinct, and no phenomenon could manifest itself.

2nd. If the causes were only liable to vary within certain limits, the phenomena which we observe would be reproduced

---

* For further information on this subject, the reader is referred to M. Quetelet's notes to M. Verhulst's translation of Sir John Herschel's *Treatise on Light.*

some day, and everything in nature would be subject to the law of periodicity. At the same time the most simple combinations of causes would give rise to periods of an immense extent.

3rd. If causes, on the contrary, have no permanent character —if the Supreme Being has permitted some to be extinguished to make room for others, periodicity is but accidental, and nothing is durable in the universe—at least, in the order of things which fall under the influence of causes—whose existence is limited.

LETTER XXX.

WHEN I spoke on the mode of action of accidental causes, I supposed that we had to measure an individual whose height remained invariable during the course of the experiments. This invariability of height might be considered as a constant cause, which should not in any way modify the value of the results.

These same results would not be sensibly altered even when the height during the course of the experiments varied about a mean, sometimes in one direction, and sometimes in another,—that is to say, without the variation being progressive in any determined direction. This character of causes is very remarkable, because examples of it are very frequent in nature. There are, in fact, very few causes which may be considered as absolutely constant.

The small variations which alter a cause, and which are only exercised within very narrow limits, may be regarded as the effects of accidental causes, added to other accidental causes which may already influence the final result. So that definitively the variable cause may be considered constant; and the accidental causes, having become more numerous and more varied, make the result sought to oscillate between wider limits of error. I suppose a piece of money thrown into the air to obtain at hazard "head" or "tail," and that after 10,000 trials there were obtained—heads 6,000 times, and tails 4,000 times. From what we have already seen, the causes favourable to heads or tails are as 6,000 to 4,000, or 3 to 2. I admit that the trials have been sufficiently numerous to have destroyed the effect of accidental causes, so that we should gain little or nothing by prolonging the trials, since we should always obtain almost identically the same ratio.

The remarkable principle of James Bernoulli consists exactly of this. That skilful geometrician had employed a part of his life in demonstrating this result, which now appears so simple to us; namely, that the mean given by a series of trials falls near the number sought within limits so much the more narrow as the trials are more multiplied. All the properties which result from his learned researches constitute one of the most honourable monuments to his memory. But Bernoulli established his calculations on the hypothesis that the number sought was fixed and determined.

It may happen that this quantity will experience small variations, such as those of which I have before spoken. Thus, in the example in question, instead of repeating the trials with one and the same piece of money (a dollar for instance), it might be made with different pieces. It is evident that the chance relative to each new coin would vary; and instead of having for either head or tail the respective probabilities of $\frac{3}{5}$ and $\frac{2}{5}$, we should have one which would vary slightly in excess or defect from the mean probabilities determined by the mode of procedure, and the substance of the coins. But the principle of Bernoulli is still applicable to this case; and has been demonstrated by M. Poisson by means of analysis. Happily, the aid of calculation is not necessary here to conceive that the small variations which we find in passing from one coin to another may be ranged among the effects of accidental causes, which are obliterated when the experiments are sufficiently multiplied. So that the results are presented as if the trials had really been made with but one coin.

In the case before us the experiments should generally be very numerous: it is for this reason that M. Poisson has designated the extension of Bernoulli's principle as *the law of great numbers.*

We have already had occasion to consider analogous examples, particularly in reference to the measurements taken on a large number of men of the same age. We then saw that all the heights collected together gave a mean, from which they individually differed according to a well-defined law, and absolutely as if an individual type had been measured a number of times by means more or less defective. This symmetry in the results does not

L

exist, nor can it exist, but in so far as the elements which give the mean can be referred to an individual type.

The measurements of the chests of the Scotch soldiers have presented us with another example of the same nature. They varied more or less from the common mean, as would vary different measurements taken on one and the same chest, but by defective means which would give rise to greater chances of error.

Your Highness will, I hope, excuse me for reverting to these two examples before quoted. I attach so much the more value to them, as they present I believe the first mathematical proof that there really exists a human mean—a human type,—at least as to size. This assertion has been strongly contested by some, and I had at that time no idea of submitting my numbers to the trials I have just indicated.

LETTER XXXI.

ON THE STUDY OF CAUSES, AND ON THE COURSE TO BE PURSUED IN OBSERVATION.
—WE PROCEED FROM GENERAL FACTS TO PARTICULAR CASES.

In statistical researches, our object is generally to discover the causes which influence social facts, and to determine the degree of their energy. These appreciations, especially the latter, sometimes become impossible; and we must confine ourselves to the study of the causes, and their tendency.

Society is not like a physical instrument which we arrange and disarrange at will, for the purpose of studying it under all its forms, in all its variations, and at the most favourable time. Woe to him who shall attempt such experiments! He would have to complain of a country which every instant changed its laws, its customs, and its relations, to find by experiment the most suitable mode of existence. We cannot then, as in the greater part of the sciences of observation, equalize at will all the influential causes, save one, so as to study the effects and modes of action of this latter. We must often proceed by other ways: we must substitute analysis for synthesis, and commence by taking the phenomenon in its most general state.

Let us remark, in the first place, that it is nearly always possible to disregard in the study of social phenomena, as in that of physical phenomena, the effects of *accidental* causes, by making the results depend on a sufficient number of observations. Thus, in studying the mortality of Belgium, I made the results depend on very large numbers, in order that we might consider the effects of accidental causes eliminated.

It is equally possible to throw out the effects of periodical variable causes, by only comparing with one another results given by an entire period, or by the corresponding parts of a period. Thus, in studying the mortality of Belgium, I should establish comparisons between the results of successive years. In default of the results of a whole year, I should compare those of many successive winters or springs, because I suppose a constant ratio between the number given by one particular season and the whole year.

If, instead of eliminating, we wish to study the effects of the period, we must operate in quite a different manner; it will be the partial numbers of the different divisions of the period that we must place before us.

The study of *variable causes which are not periodical* is more difficult. The best thing to be done, in such a case, is to compare the result which is supposed to be influenced with many other similar results obtained for other years, so as first to judge whether there are any anomalies or sensible variations in these results. When a difference is found, we must discover whether it is periodical or fortuitous. We must next compare the result which we suppose to be influenced with the causes which may have made it vary. For example, I have shown in one of my preceding letters that the mortality in the years 1835 and 1836 was less than during the contiguous years. I was led to think that this diminution in the deaths was owing to the price of provisions; and to assure myself of this, I compared the returns of the deaths for the ten years from 1831 to 1840 with the prices of wheat and rye. I found that, in fact, the progress of these two statistical elements was nearly parallel. I concluded from this that the variable causes which produced a diminution of mortality was probably due to the fall in the prices of grain.

When we wish to study *constant* causes, we must commence by eliminating the influence of accidental causes, and have regard to the variable causes, to clear also of their effects the results on which we work. It is for these reasons that the researches should rather be carried out on years which are complete, and not anomalous.

The surest course then will be to study the fact first in its greatest generality,—that is to say, under the influence of all causes constant and variable; then of all these causes less one,—that one whose existence and tendency we wish to discover. For this purpose we shall partially group the numbers which have concurred to give the general result, collecting them in the manner most advantageous for showing the influence we suspect, and its mode of action.

I will endeavour to illustrate by an example that which may be too abstract in theory. Only I pray your Highness not to be alarmed at the sight of the figures which I shall be forced to employ,—they are in this case my forced auxiliaries. But I will not lose sight of the fact that they are always more or less misplaced on the ground on which I shall make them interfere: it is enough to say that I will use them soberly.

An interesting question which connects itself with many branches of our knowledge, but which above all touches the dearest affections of man, is that of the stillborn. If this question has not yet been examined with all the attention it deserves, it is not because it has been lost sight of by men of science or by statesmen: the mysteries which surround it are rather owing to the deficiency of good observations.

The defects which exist in this respect are partly owing to the fact that the declarations required by law are either not made at all, or are only made in a very irregular manner. Stillborn children are besides often confounded with abortions. The law itself is not sufficiently precise; besides, in a great number of civilized countries the laws preserve the strictest silence on the subject of the stillborn.

For some years the Belgian Government has taken especial care to procure uniformity and exactness in this interesting part of statistics: the information, which has been collected during the last three years, already furnishes us with more exact notions than we have hitherto possessed. I will begin by showing the general numbers.

	TOWNS.			RURAL DISTRICTS.			
YEAR	Male.	Female.	Ratio.	Male.	Female.	Ratio.	TOTAL.
1841	1220	949	1·29	1976	1387	1·43	5532
1842	1205	915	1·32	1939	1415	1·37	5474
1843	1296	934	1·39	2073	1456	1·42	5759
MEAN . .	1240	933	1·33	1996	1419	1·41	5588

NUMBERS OF STILLBORN OF BOTH SEXES.

Belgium then has numbered annually 5,588 stillborn children. I think we may accept this number as sufficiently free from the effects of such small accidental causes as might affect its value; in fact, it differs very little from the numbers given by each year separately.

The observations made in the rural districts, and even those in the towns, appear to be sufficiently numerous for the effects of accidental causes to be considered as eliminated in the results. This remark even extends to the numbers which establish the distinction of the sexes.

This fixity in the number of stillborn is a first fact which appears to me very remarkable: it already tends to show that it is the result of well-determined causes.

Are these causes constant or variable? During the three years which have just passed, they seem to have acted in a constant manner. The different numbers, however, of 1843 are the greatest, for whatever distinction we establish. This increase (small, indeed) should have a cause. I think it is to be found in the care that was taken during this year to give the number of stillborn with more exactness than before.

Your Highness will remark (and this is an essential point) that I am far from regarding the mean 5,588 as exactly representing the annual number of stillborn, although I admit that it is not sensibly influenced by accidental causes. I think, on the contrary, that this number is too small, on account of concealments and neglect in registering.

But omissions in the affairs of administration are not made fortuitously. We may regard them as constant causes, or at least as variable causes which exert a definite influence on the general results. Is it not known that the Post Office receives each year nearly the same number of letters unsealed, or with insufficient addresses? Do we not see the same regularity in the offences against the police regulations?

I will only take the mean 5,588 then as representing approximately the absolute number of stillborn. By comparing it with the number of the population, which was at the end of 1842 4,172,664, I find annually 1 stillborn to 747 inhabitants.

Another remark, which seems to me very interesting, here presents itself: it is, that although the numbers of stillborn are very defective in an absolute sense, they may nevertheless give very correct ratios when compared with one another. To explain: if we wished to know the influence of sex on the stillborn, it would matter little if the two numbers compared were each too small or too great,—it is sufficient to be convinced that both vary in the same ratio; and then the result of the comparison is freed from this source of error. But everything leads us to suppose that concealments and omissions to register are made without reference to sex.

When we compare the male and female stillborn, the ratio is 1·33 to 1 in towns and 1·41 to 1 in the rural districts. The stillbirths of males are comparatively more numerous; and the inequality is always a little less in towns than in the country. Is this merely accidental? To ascertain whether it is, we must consult the observations of a greater number of years.

From the observations of the three years before mentioned, it would be difficult to arrive at a conclusion, because the smallness of the numbers exercises too great an influence on the variations of the ratios.

Let us remark, in passing, that observations which are sufficiently numerous to determine the action of a certain kind of causes may not be sufficient to discover with an equal degree of precision actions of another nature.

We may appreciate in a surer manner, and by another mode, whether the difference of residence in towns or in the country has caused a variation in the number of stillborn.

In the research I propose to make, I will take the number born alive as a term of comparison. This last element is one of the most precise of the state returns, at least in Belgium. I will compare then the number of infants born alive and the number of stillbirths in town and country.

The following table may give rise to melancholy reflections; in fact, it will be seen that the number of stillborn in towns is, in comparison with the number in the country, nearly double; and this difference is not accidental, but re-occurs each year with nearly the same value.

RATIO OF LIVING BIRTHS TO STILLBIRTHS IN BELGIUM.						
YEAR.	IN TOWNS.			IN THE RURAL DISTRICTS.		
	Infants born alive.	Stillborn.	Ratio.	Infants born alive.	Stillborn.	Ratio.
1841	35,053	2169	16·2	103,082	3363	30·6
1842	35,156	2120	16·6	99,871	3354	29·8
1843	34,817	2230	15·6	98,094	3529	28·0
MEAN . .	35,009	2173	16·1	100,349	3415	29·4

What may be the cause of this afflicting tribute levied on towns?
Does it proceed from a greater negligence in the country in registering the stillborn? I think that there may exist some inequality
on this head; but it certainly would not be sufficient. Should we
attribute it to the dissipation and depravity of towns, or to the
constraints of the clothing? I abstain from pronouncing.

Moral causes, perhaps, exercise a greater influence on this inequality than physical causes. I will give an example of this, in
the comparison I am about to make between children born alive
and stillborn, distinguishing legitimate from illegitimate births.
The following is what we learn from statistical documents:—

RATIO OF LIVING BIRTHS TO STILLBIRTHS IN BELGIUM.						
YEAR.	LEGITIMATE.			ILLEGITIMATE.		
	Infants born alive.	Stillborn.	Ratio.	Infants born alive.	Stillborn.	Ratio.
1841	128,781	4991	25·8	9354	541	17·3
1842	125,841	4885	25·7	9186	591	15·5
1843	123,603	5125	24·1	9308	634	14·7
MEAN . .	126,075	5000	25·2	9283	589	15·0

There is then a great difference in the number of stillborn, taking into consideration the legitimacy of births: the ratio is about 25 to 15, or 5 to 3. Unfortunately, we cannot doubt—and the figures I have just cited support the belief—that immorality and misery are destructive causes which affect man even before he has seen the light.

The preceding result may explain to us, at least in part, the difference we have noticed between towns and the country in the ratio of stillborn. We know that in towns the illegitimate births are much more numerous than in the country: this ratio, all other things equal, is annually 23 to 7. Consequently the stillbirths ought also to be more considerable.

In the different comparisons I have just established, I have taken care to eliminate the influence of annual periodical causes by only comparing together the results given by a whole year. If I wished to determine the influence of these causes, I should, on the contrary, compare the results of each month individually. The following table exhibits these comparisons. The first three columns show, month by month, the number of stillborn registered in Belgium during the years 1841, 1842, and 1843. These data are extracted from official publications *sur le Mouvement de l'Etat Civil*, which the Minister of the Interior makes every year.

The numbers of each year individually demonstrate that there are fewer stillborn in summer, and that on the contrary there are more in December and the following months; but we can only indistinctly perceive a law connecting these numbers together. This inconvenience is principally owing to two causes,—

1st. The observations are not sufficiently numerous.

2nd. The months are of unequal length.

But to guard against the influence of these two causes, I have added together the numbers furnished by the three years, and placed the sums in the last column but one of the table. These numbers enabled me to calculate the numbers in the last column, where each month is supposed to consist of thirty days.

This last column clearly indicates a periodicity: the greatest number of stillborn is found in the month of March. This number diminishes until June, which presents a *minimum;* and it increases afterwards, until it attains the *maximum* in March.

| NUMBER OF STILLBORN PER MONTH IN BELGIUM. | | | | |
Months.	1841	1842	1843	Total.	Each 30 Days.
January . .	481	541	548	1570	507
February .	467	458	499	1424	508
March . .	506	550	559	1615	521 *max.*
April . . .	504	489	487	1480	493
May . . .	472	443	479	1394	450
June . . .	392	401	386	1179	393 *min.*
July . . .	430	422	418	1270	410
August . .	435	408	458	1301	420
September .	439	399	439	1277	426
October . .	437	432	485	1354	436
November .	461	417	476	1354	451
December .	508	514	525	1547	499
THE YEAR	5532	5474	5759	16765	460

The law of mortality is here nearly the same as for an infant after birth. I have elsewhere shown (and Messrs. Edwards and Villermé had discovered it before me) that the heat of summer causes a slight increase of mortality among the new-born; so that the *minimum*, which for other ages generally happens in July, is hidden (or rather displaced), and occurs in the month of June.

Excess of cold, as well as of heat, is fatal to the infant at the time of its birth. Excess of cold especially exercises its fatal influence over our whole life. Singular condition of the human race! our frail existence is submitted to all the vicissitudes of temperature, and even the bosom of a mother cannot shelter from death.

All the preceding results fully confirm those I had already arrived at in my *Essai de Physique Sociale* before 1835, and those obtained by Dr. Casper for the Prussian Kingdom; only they determine more precisely the values of the ratios calculated from the little information I have been able to collect. The influence of sex—that of towns and country, that of legitimacy and illegitimacy, and, lastly, that of the seasons—preserve nearly the same value in their determinations. I mention these circumstances because they tend to prove that the preceding results are not peculiar to Belgium, but that they possess a character of generality, which makes them more important.

LETTER XXXII.

27th March, 1845.

THIS morning I looked with sadness on my garden. Through
my frost-covered windows, I saw the shrubs bent down with the
weight of snow. Nature, cold and dead, seemed wrapped in a vast
winding-sheet. I recollect, however, that at the same season I
have many times seen the first flowers of spring opening their
corollas, and verdure adorn these mossy grounds now hardened
by the ice.

More fortunate than Belgium, Italy has doubtless already seen
the crocus and the cowslip blow. That ocean of verdure and of
flowers, with which she is covered, will progressively extend its
waves to our regions; for, notwithstanding the causes which fetter
it, its march is as much regulated as is that of this other ocean
which bathes our coasts.

Even the flower, so modest as it is, is not the product of a
caprice of Nature: its frail tissue requires favourable conditions
for its development; and to arrive at its full bloom, it follows laws
as stable as those which preside over the progress of worlds.

If our eyes were sufficiently penetrating to embrace the succession of phenomena which are accomplished—I do not say in one
plant, but the entire collection of plants—in all the vegetation
which displays itself periodically on the surface of our planet, we
should doubtless be astonished at the harmony and regularity with
which these same phenomena are reproduced. This study has
always charmed me; and if my knowledge has not permitted me
to find the laws which govern the great metamorphoses to which
our globe is subject, I have attempted at least to discover the

relations which exist between some of the phenomena, and the principal causes on which they depend.

This study is not new,—the celebrated Linnæus, in particular, has applied to it all the powers of his intelligence, and of his brilliant imagination. He understood that, in order to obtain the phenomenon in its integrity, it must be studied simultaneously at a great number of points, and he appealed to his friends; but it is difficult to obtain from men a concurrence of labours which should extend over a long period, and whose fruits would be tardy. Linnæus did what might have been expected from so powerful an organization : he produced the calendar and the timepiece of Flora; but he left still enveloped in darkness the mysterious bonds which exist between the different natural periods of plants, and the elements which act most directly in giving birth to them.

We know that blooming, for instance, is much influenced by temperature, and that the phenomenon also depends on the nature, the age, the exposure of the plant, the quantity of light, the moisture of the air, the state of the heavens, the winds, and an infinity of other causes; but we have not yet succeeded in assigning to each of these causes its individual influence. This research still remains to be made.

The interest which this subject possesses will doubtless serve as my excuse, if I try what course is to be pursued, to direct one's self with some prospect of success across this labyrinth where so many threads cross each other. It is, moreover, one of the most beautiful examples I can select for the development of my ideas.

The essential point is first to discover what causes may influence the phenomenon of blooming. Were I required to make a complete enumeration, I should perhaps have to renounce entirely this kind of research; but I can here confine myself to the consideration of the predominant causes, as is the practice in physical sciences, returning afterwards to the secondary causes, which I may think may be neglected in a first inquiry.

In a preceding letter I have proceeded by analysis to arrive at a knowledge of the truth : I think it will be preferable here to employ synthesis, and to study the phenomenon under its most simple form, so as to rise gradually to the general case.

I range in four principal classes the causes which may influence the blooming of plants.

Geographical causes, such as the latitude, the longitude, and the altitude.

Local causes, such as the nature of the soil, the exposure, and the quantity of light.

Individual causes, such as the age and vigour of the plant.

Meteorological causes, such as the temperature, the nature of the winds, the moisture of the atmosphere, the quantity of rain, and the state of the heavens.

It would be very difficult to study all these causes simultaneously, and to discover the portion of influence due to each. The most sure way will be to proceed from simple to compound.

Let us first remark that we can eliminate the effects of causes which belong to the three first categories by always observing the same plants, and on the same point of the globe, exposed in the same way, and in the same ground : the differences which will present themselves in the epochs of putting forth the leaves, of blooming, of the ripening of fruit, or of the fall of the leaves, when all other things have been made equal, can only proceed from meteorological causes.

I will suppose, moreover, that it is the same observer who traces the development of plants, and takes care to register their natural epochs. Many observers would no doubt observe in different ways, and one would have a propensity to mark the epoch of blooming later than another. This difference, which we might call the *personal equation of observers,* would be a new cause of error, which we must take into consideration.

I will admit that every precaution has been taken to render as equal as possible all the general causes, except the meteorological causes which may make a difference in the phenomena of vegetation. These conditions will not be difficult to fulfil, since it will be sufficient for me to observe the same plants, in the same place, and similarly exposed during many successive years, in order to eliminate by repetition the effects of fortuitous causes. I must take care at the same time to register the meteorological state of the atmosphere whose influence I wish to determine.

It is very evident, in the first place, that if the atmospheric circumstances remained every year exactly the same, the plants would be covered with leaves, would flower, produce their fruit, and lose their leaves, each respectively at the same period of the year, since the causes remaining the same their effects would be invariable; but since the condition of the atmosphere undergoes very sensible modifications from year to year, the natural epochs of plants will undergo the same. It is these modifications that we will first study.

When we have carried this study as far as possible, and have determined the share of each of the meteorological causes, we shall be prepared to establish for each plant the day of putting forth its leaves, of flowering, &c., by supposing a normal year in the locality in which our observations have been made. In other words, we can eliminate the effects of meteorological causes.

If, instead of confining ourselves to one only of each kind, we had taken many, placed in different ground and in different situations, we might have determined the influeuce of *local causes*. The difference of exposure to the sun or shade, to the North or South, in a dry or damp ground, would have produced an inequality in the period of flowering of each plant. This inequality, repeated each year, furnishes us with means of determining the local causes.

I have supposed that all the plants which served for the preceding observations were equally healthy, equally aged, and equally vigorous, so that all the *individual causes* were the same; but supposing these conditions do not exist, we may determine the influence of individual causes. We should observe in the same ground, in similar position, and in precisely similar circumstances, plants of the same species, differing only in age, or in vigour, or in some other character.

The study of geographical causes is quite as easy in theory; but it presents great difficulties in practice,—first, because different observers must be employed, who rarely observe in the same manner. Again, it would be nearly impossible to observe with individuals perfectly alike in age and strength, in order to find the same soil and the same exposure. We must then, by previous calculation, reckon the meteorological causes, local and individual, to obtain results which may be compared one with another, or which at least are not influenced by geographical causes.

My method of proceeding here differs from that indicated in my preceding letters. I commence by observing the fact in its most simple form, by endeavouring to make all the other influencing causes but those I wish to examine equal to one another. Then, having discovered the influence of the latter, I eliminate their effects in the ulterior observations, or I make unequal in their turn those causes which I had made equal at first. Proceeding thus from simple to compound, I finally determine the influence of those causes which it does not depend upon me to render equal, such as the altitude, the latitude, the locality, or distance from the sea, &c.

In statistical facts on the contrary, as we have already seen, we generally study the question first in its generality, and descend from compound to simple.

But I do not wish to occupy myself with generalities in that which concerns the study of the natural epochs of plants. This study is too interesting for us not to endeavour to enter upon it, at least, by its most accessible points. I do not deceive myself as to the difficulties it presents, but I think they may be surmounted. I will occupy another letter with the applications that may be made of the method of observation which I have endeavoured to sketch.

LETTER XXXIII.

THE STUDY OF CAUSES CONTINUED.—BLOOMING.

I NOW propose to fulfil the engagement I made at the close of my last letter.

I will commence by reducing the question to its most simple form. I will suppose that it is wished to determine the time of blooming of some well-known plant, whose species cannot be mistaken. This condition is essential, especially if we wish to make comparisons. We must choose a plant which grows all over Europe, and which is easily met with.

In these respects the lilac (*Syringa vulgaris*) will perfectly answer my views. It is the universal ornament of gardens; and in the country it is rarely absent from even the most modest houses.

When curiosity and taste for Botany led me to think of the phenomenon of flowering, I perceived that the lilac above all claimed my attention. I also perceived that I could not leave it to others to take notes, which probably would not have been comparable with mine. I mark as the day of flowering that on which the first corolla opens and shows the stamina, unless the opening is an extraordinary case, which the state of the plant will sufficiently show.

The place in which I observe is the garden of the Observatory: its extent, and its distance from the nearest houses, render it very favourable for this kind of observation. The ground (generally pretty dry) is nearly at the top of the hill on which Brussels stands, at a height of about 194 feet above the level of the sea.

These different details are necessary, if we wish to take account of all the causes which may influence the phenomenon. I should add that lilacs are very numerous in the gardens of the Observatory; and that their age may, on the average, be perhaps twelve

M

years. Those which receive the rays of the sun more directly are also the first to be covered with flowers. In other respects, I have not remarked that any part of the garden was less forward than the other, excepting the part immediately to the north of the building. Generally the flowering took place on the same day, on many plants at once.

My first observations on vegetation date from the commencement of 1839; so that I have already been six times enabled to observe the blooming of the lilac, and to register the dates on which the first flowers have opened. The following table gives the results obtained by observation on this plant:—

Time of the Flowering of Lilacs at Brussels.

1839	.	.	.	.	10th May.
1840	.	.	.	.	28th April.
1841	.	.	.	.	24th „
1842	.	.	.	.	28th „
1843	.	.	.	.	20th „
1844	.	.	.	.	25th „

MEAN . 27·5 April.

It is then from the 27th to the 28th of April that the lilac blows in Brussels. In 1839 it was retarded twelve days; in 1843, on the contrary, it was eight days in advance: the difference between these two extreme periods is twenty days. This determination allows of some uncertainty : I estimate it as two or three days at most for the lilac.

Now what are the causes which have produced the differences which we have remarked? These causes can only be deduced from meteorological circumstances, since all the other circumstances which generally modify vegetation were the same.

Among meteorological influences, it is agreed to place in the first rank that of the temperature. It is thus that observers have first directed their attention to the study of this cause. Reaumur, I believe, first thought of taking the sum of temperatures, with a view to ascertain the day on which a vegetable phenomenon should occur.* The Abbé Cotte followed this idea; and in his opinion

* *Mémoires de l'Académie des Sciences,* 1735, page 559 ; and the *Traité de Météorologie,* by P. Cotte, page 424.

the opening of a flower was the result of the temperatures to which the plant had been previously subject. In other respects, the date which served him for a starting point was quite artificial; he reckoned from the 1st of April. It is sufficient to transport ourselves in thought to the southern hemisphere, in order to see how small a real value this epoch has.

Another equally well founded objection may be made to this mode of reckoning,—it is that all the degrees of the thermometer have the same value, whether they refer to hot or cold days. Three days of a temperature of 8° centigrade (46° Fahrenheit) in the month of June, for example, should produce the same effect as one day of 24° centigrade (108° Fahrenheit). But these three cold days, following one another at the time when the sap is in all its activity, will rather retard vegetation; whilst a temperature of 24° (centigrade) will give it a new force, and will cause a great number of flowers to open.

Notwithstanding the objections that may be made to it, Reaumur's idea is ingenious, and doubtless varies but little from the truth. This skilful physician was right in ascribing to heat the greatest influence in the phenomenon of vegetation, and in endeavouring in his first research to determine this influencing cause, leaving until afterwards the discovery of other causes and the appreciation of their effects.

It is then to the determination of the influence of temperatures on blossoming that we must first direct our attention. For this purpose, we will try to substitute for the starting point of the Abbé Cotte a natural epoch,—that is, an epoch marked by Nature herself. After many essays, I have concluded that we should select the instant when the plant comes out of its winter sleep, and when the sap begins to flow. I fix this instant at the end of the winter frosts, and in the following manner for the six years from 1839 to 1844:—

In 1839	.	.	.	. the 14th March.
1840	.	.	.	3rd „
1841	.	.	.	2nd „
1842	.	.	.	27th January.
1843	.	.	.	25th „
1844	.	.	.	25th „

It sometimes happens that severe frosts return after the sap has begun to circulate, and destroy the small leaves and flowers which were developed. If these frosts be prolonged, plants enter again their winter sleep, and the starting point is retarded.

This point deserves to be fixed with the greatest care. I will presently show a method of defining it, perhaps more exactly than by the cessation of continued frosts.

I have already stated what leads me to regard as faulty the proceeding by which Cotte and the botanists who have followed him calculate the epochs of blossoming. The force exercised by the temperature is of the same nature as actual force. It is by the sum of the squares of the degrees, and not by the simple sum of the degrees, that we must appreciate its action. To put my conjecture to a trial, I have formed two tables, for each of the years of observation,—the one containing the sum of the temperatures, and the other the sum of the squares of the same temperatures from the fixed period which has been previously discussed. I have found, on comparing these two tables, that on the average 462° (centigrade) of temperature are required to produce the flowering of the lilac according to the ideas of Cotte, and a total of 4,264, as the sum of the squares of the several degrees, according to my method of reckoning.

Admitting these numbers, and seeking the dates to which they correspond in my two tables, calculated as I have just stated, I find as the epoch of flowering—

EPOCHS.	ACCORDING TO		ACCORDING TO OBSERVATION.
	Temperature.	Squares of the Temperature.	
1839	10·5 May	9·3 May	10 May
1840	4·0 ,,	2·2 ,,	28 April
1841	23·5 April	23·0 April	24 ,,
1842	22·5 ,,	27·3 ,,	28 ,,
1843	19·5 ,,	19·7 ,,	20 ,,
1844	22·0 ,,	23·5 ,,	25 ,,
MEAN . .	27·0 April	27·5 April	27·5 April

These numbers only agree with the numbers observed so far as to leave the question undecided. However, the method I propose gives more satisfactory values, especially for the year 1842. As one single plant could not resolve the difficulty, my observations have necessarily been extended to a great number of plants; and I think that the total results can leave no doubt.

The greatest variations, calculating by the second method, are generally within the limits of probable error.

Admitting the number 4,264 as that which corresponds to the epoch of the flowering of the lilac, we may by subtracting the squares of the temperatures of each preceding day arrive at the epoch of revival.

If plants, in order to blossom, were only subject to the action of the temperature, we should attach to each a number analogous to 4,264, which would determine the instant of its flowering, whether in a greenhouse or in any region whatever of the globe. This would be a *constant cause*.

Let us suppose that we have in fact determined the mode of action of the temperature, and that we could reckon its influence, we should know what correction each of the dates above enumerated should undergo. But if the temperature were the sole influencing cause, after the correction was made, the dates should become identically the same, or at least should only differ within the limits which relate to the uncertainties of observation. Admitting that they were actually within these limits, it would become very difficult, without a great number of observations, to determine the actions of other influencing causes, since these actions would lose themselves in the effects of accidental causes.

If, on the contrary, after having submitted the dates to the corrections required for temperature, we yet find differences which cannot be justified by the errors of the observations, we must seek to explain them by the influence of winds, the quantity of rain, the state of the heavens, or other meteorological causes. But in making use of the observations that I have been able to collect, I find that the correction for the temperature alone is sufficient to bring the calculated period of flowering within limits which include the uncertainties of observation: it will be useless then at present to seek to proceed further.

It is very easy to take account of *local causes*. It is sufficient to place, in different soils and positions, plants of the same species, the same age, and as similar as possible, and to observe their development simultaneously.

The same with respect to *individual causes*. We must place in the same ground, and in the same position, plants of the same kind, but different in age, variety, and vigour.

The influence of these different causes may be determined in so much more certain a manner, as the errors of the personal equation are unmixed with it.

This is no longer the case when we wish to determine the influence of *geographical causes*: the phenomenon is complicated by the impossibility of separating the different sources of error. Thus, in one and the same town, we may obtain very different results, in our endeavours to determine the epoch of the blooming of a plant. We must necessarily add together the effects proceeding from the difference of the soil, of the position, of the nature of the plant, of the manner of observing, &c. If the distances are great, the difference of the meteorological causes is added to the difference of latitudes and heights.

Notwithstanding these numerous difficulties, the hope of collecting some information on this interesting question has led me to undertake an enterprise, to which I have already referred in one of my previous letters. I might have been dispirited, however, by the thought that Linnæus, the greatest promoter of Botany, had nearly failed in a similar enterprise. The illustrious and learned Swede, in fact, invited his friends to observe with him the blossoming of plants; but the results which he gives in *Aménités Académiques* will not allow of any conclusion being drawn from them: probably he felt this himself, for he abandoned this kind of research. Perhaps this great botanist, notwithstanding his penetration and his correctness of judgment, had not made his question sufficiently precise, had not determined the species and variety which should be observed, or the instant when the observation should be taken. We find in the numbers which he gives inexplicable discordances.

In following the steps of this great master, and availing myself of the counsel of men learned in these matters, I do not flatter myself that I shall be able to resolve so difficult a problem; but I hope, however, that I shall not entirely fail, if those who second me will persevere in their efforts. I collected by myself observations for two years before asking foreign assistance: my observations began in 1839. Two years afterwards I requested some Belgian observers to join their efforts to mine; and it was not till 1842 that I dared to make an appeal to the learned of other countries. I have thus collected observations at many points in Holland, England, France, Switzerland, Italy, and Germany. In 1840 M. Fritsch conceived the same idea, and commenced collecting at Prague observations on the natural epochs of plants, which he consigned to the meteorological collection of M. Kreil. Analogous observations have also been made in the United States.*

Drawing from these different sources, I have succeeded in forming the following table, in which will be found the epochs of the flowering of the lilac in the different years, and also the *mean epochs* calculated in comparison with the observations at Brussels.†

These observations are still too few for the results to be accepted with confidence: many dates are only determined from one year's observation. We should have patience to collect the notes of many consecutive years, so as to eliminate from the general results everything accidental in the particular determinations.

For want of time we must have recourse to a greater number of plants, and operate on each as I have done with the lilac. Each plant thus furnishes a particular date; and the whole of these dates, compared with those obtained at Brussels, show the advance or retardation of blooming in different localities with so much the more precision as the number of plants which have combined to determine it have been greater.

---

* The observations of the State of New York are taken from the annual reports of the university professors. The numbers which I use are deduced from the results given by 8 towns in 1839, by 15 in 1840, 9 in 1841, 11 in 1842, and 10 in 1843.

† The mean epochs are calculated in the following manner. The lilac flowered at Parma in 1843 on the 10th April, that is ten days earlier than at Brussels; in 1844 it flowered eight days earlier than at Brussels: so that the mean advance was nine days for these two single observations. But the lilac blossoms at Brussels on the 28th April: we may therefore conclude that the date of flowering in Parma is the 19th April.

EPOCH OF THE BLOSSOMING OF THE LILAC.							
PLACE OF OBSERVATION.	1839	1840	1841	1842	1843	1844	MEAN EPOCH.
Parma . .	...	...	...	...	10 April	17 April	19 April
Venice . .	...	...	...	...	...	16 ″	19 ″
Paris . . .	...	...	...	20 April	...	...	21 ″
Brussels . .	10 May	28 April	24 April	28 ″	20 April	25 April	28 ″
Liége . . .	...	...	25 ″	...	...	...	29 ″
Louvain . .	...	...	29 ″	27 April	...	...	30 ″
Ghent . .	...	...	28 ″	30 ″	23 April	27 April	1 May
Bruges . .	...	...	...	...	26 ″	...	4 ″
Ostend . .	...	...	...	10 May	26 ″	...	7 ″
Utrecht . .	...	...	...	6 ″	...	6 May	7 ″
Joppa, Deventer }	...	...	...	...	...	5 ″	8 ″
Lochem, Gueldre }	...	...	...	...	...	7 ″	10 ″
Gromingen, Frislland }	...	...	...	12 May	...	...	12 ″
Prague . .	...	8 May	8 May	8 ″	3 May	...	10 ″
Environs of Cambridge }	...	...	...	...	1 ″	...	9 ″
Munich . .	...	...	...	2 May	6 ″	4 May	10 ″
State of New York }	16 May	16 May	26 May	16 ″	22 ″	...	21 ″

The following table will give an example of the research spoken of in the preceding page. I will only remark that the difference of dates for Geneva, Alaix, and the environs of London are not the product of simultaneous observations.

DIFFERENCE IN TIME OF BLOSSOMING REFERRED TO BRUSSELS.						
PLACES.	Syringa Vulgaris.	Sambucus Nigra.	Aesculus Hippoc.	Philadelphus Coronarius.	Digitalis Purpurea.	MEAN.
	DAYS.	DAYS.	DAYS.	DAYS.	DAYS.	DAYS.
Parma . . .	— 9	— 10	— 8	— 11	— 14	— 10
Venice . . .	— 9	— 2	...	— 14	...	— 8
Alais	— 44	— 16	— 16	— 22	...	— 24
Geneva . . .	— 2	+ 8	— 5	...	...	0
Paris	— 7	...	...	...	...	— 7
Valognes . . .	+ 7	— 4	...	...	— 8	— 2
Liége . . .	+ 1	...	...	+ 3	— 2	0
Ghent . . .	+ 3	— 2	+ 1	+ 2	— 5	0
Bruges . . .	+ 6	+ 3	...	...	...	+ 3
Ostend . . .	+ 9	+ 10	+ 9	+ 7	+ 1	+ 7
Environs of London	+ 8	...	+ 5	...	+ 15	+ 9
Do. of Cambridge	+ 11	+ 10	+ 11	+ 9	...	+ 10
Vucht (S.° Brabant)	...	...	+ 9	— 6	...	+ 2
Utrecht . . .	+ 9	+ 12	+ 8	+ 12	+ 9	+ 10
Lochen . . .	+ 12	...	+ 6	...	...	+ 9
Joppa, Deventer .	+ 10	...	+ 7	...	...	+ 9
Groningen . .	+ 14	+ 13	+ 17	+ 19	+ 7	+ 14
Prague . . .	+ 12	+ 11	+ 7	+ 15	...	+ 11
Munich . . .	+ 10	+ 23	+ 15	+ 32	+ 8	+ 18
Jever . . .	+ 26	+ 29	+ 14	...	+ 33	+ 25
Rochester, U. S. .	+ 16	...	+ 17	...	+ 27	+ 20
Epoch of Blossom-ing at Brussels .	28 April	22 April	1 May	19 May	4 June	

Although the advance or retardation of blooming indicated by each plant in a given locality is not exactly the same, it may however be seen that this element varies between not very wide limits, considering the numerous causes of error by which the phenomenon may be influenced. The numbers given by the localities of which Brussels is the centre especially deserve greater confidence, on account of their being simultaneous, and of the attention that is paid to get as nearly as possible the same phase of blooming,—that is, the instant that the corolla opens. The learned who have been engaged on the subject of blooming have not always taken the same precautions; and they have more generally referred the phenomenon to the time when the plant was more or less covered with flowers. Thus their annotations when compared with ours are always some days later.

I shall now seek to discover, by the aid of the elements in our possession, the influence which the difference of latitudes and of elevations exercise on blossoming. In this point of view the phenomenon has already been studied by many learned men.

Of all the towns contained in our catalogue (with the exception of Rochester, United States,) that which differs most in latitude from Brussels is Alaix, in the department of Gard: the difference is 6° 46ˢ. Its elevation is 270 feet greater than that of Brussels, and its blossoming ought consequently to be retarded a little. The retardation is nearly 4 days, as we shall soon see. These 4 days added to the 24 indicated in the table give a retardation of 28 days for 6°46ˢ, or nearly 4·1 days for a difference of one degree of latitude. This determination agrees with the two following.

The time of blossoming of the lilac, the elm, the birch, the linden, and the oak is known for the environs of Naples.* Comparing the dates with those of Brussels, we find that in the latter place the blossoming took place 36 days later; and as between Naples and Brussels the difference of latitude is nearly 10°, it can be seen that we must reckon 3·6 for one degree of difference in latitude. Let us add that the environs of Naples differ little in elevation from those of Brussels.

The illustrious Linnæus, in his *Aménités Académiques*, gives the

* *Almanach de Berghaus*, 1840, page 63.

epoch of blossoming of the gooseberry, the willow, the birch, and the poplar for the locality of Grippenberg in Lapland. These epochs compared with those of Brussels give for the latter place an advance of 48 days. Considering that the difference of the latitudes is about 15°, I find 3·2 days advance for one degree. I am not aware at what elevation the observations were made.

We may then suppose that we are not far from the truth, in admitting that blossoming in general is subject to a retardation of 34 days in advancing 10° towards the North. M. Berghaus, in his Almanack for 1840, published at Gotha, estimates this advance at 40 days for the temperate zones. It is less in the North. He only made it 34 days between Hamburgh and Upsal in Sweden, which agrees with the preceding determination; whilst he found it to be 74 days when the south of Germany was compared with the environs of Smyrna in Asia Minor. He also estimates that in our climates we must reckon a retardation of from 10 to 14 days in vegetation for an elevation of 1,000 feet. Our preceding table will afford us the means of verifying this appreciation.

The highest point above the level of the sea, where observations have been made comparatively with Brussels, is Munich, the elevation of which is estimated at 1,726 feet. This height surpasses that of Brussels by 1,532 feet. But blossoming was observed in the latter town 18 days earlier than at Munich; although we might, on the contrary, have expected a retardation of 10 days, in consequence of the difference in latitude between the two towns, which is 2° 42ˢ. The retardation of Munich, taking the last correction into account, is then really 28 days for 1,532 feet, or nearly 6 days for 100 yards in height.

Next to Munich, Geneva is the most elevated point in comparison with Brussels : the difference of height is 1,141 feet, and of the latitude 4° 39ˢ. In consequence of this last difference, blossoming should occur at Geneva 17 days earlier than at Brussels; whilst it really occurs on the same day, according to the few observations I have been able to collect. The retardation produced by the elevation of the town above the level of the sea compensates, then, for the advance which should arise from the difference of latitudes. But 17 days' retardation in a height of 1,141 feet gives a retardation of nearly 5 days for every 100 yards. This determination is a little less than the preceding, but yet superior to that given by Berghaus: I think, however, that it is not exaggerated.

The two corrections which have just been calculated, for the purpose of judging the effect of the difference of latitudes and heights in the phenomenon of blooming, may be useful when comparing localities not far distant from one another, but not so when considering a great extent of country. The data cited by M. Berghaus already prove how cautious we must be in this respect. These differences arise from our not being placed in the true point of view for analyzing the phenomenon.

We have discovered that the element really influential on vegetation is the temperature; but in order that the generally adopted method which I have previously followed in my calculations may be available, the isothermal lines should be parallel, which as we well know is not the case. Even the isothermal lines are not sufficient to resolve the problem. The plants which develop themselves and flourish in spring time depend only on the warmth they receive during the season, and not on the temperature of the rest of the year. We are here obliged to revert to our first estimate, and to say that to each plant is attached a constant, the square of a certain number of degrees of warmth necessary for the occurrence of the phenomenon. Whether the plant is found in such-and-such a latitude, at such-and-such a height, in the open air or in a greenhouse, it is the temperature that must be considered. Thus are explained all the anomalies which present themselves in this kind of research.

Geographical causes have no influence but by the variations they cause in the temperature. It is useless then to take them into consideration, if we regard the thermometric state of the place where we observe the blooming. I wish to know, for instance, at what time the lilac flowers at Prague? According to my method of calculating, I should commence by seeking the time when the frosts of winter cease in this town: then I should take the sum of the squares of the mean temperature of each day from that date, until I formed the number 4,264. This last epoch will be that of flowering.

Following that course which regards latitudes and heights, we should, in comparing Prague with Brussels, calculate that 0° 46ˢ the difference of latitude gives an advance of 3 days; and that, on the other hand, a difference of 384 feet gives a retardation of 6 days: so that the retardation for Prague in comparison with Brussels would be 3 days. Instead of this, we find a retardation of 11 days.

I think it right to remark that the following consequences result from my mode of calculating blossoming.

1st. The line which may be drawn on the globe, through all the places where a plant may flower on the same day, is not necessarily parallel to the line which passes through the places where this plant flowered ten or twenty days earlier.

2nd. The *isanthesic* lines — that is, the lines of simultaneous flowering—have no general character, but vary at different periods of the year, and tend to approach the isothermal lines.

3rd. The time which elapses between two successive phases of the same plant may not be the same in different parts of the globe. If in England, for example, fifteen days separate the epoch of the putting forth the leaves from that of the flowering of the lilac, this interval would not be so long in Italy or Spain, where the temperature increases more rapidly.

I must here pause, on account of the length this letter has attained. However, I feel that I have yet many remarks to make to dispose of the objections which may reasonably be made. I did not intend to examine thoroughly the so interesting subject of the periodical epochs of plants : I shall make it the subject of a separate work. I have only wished to present one example of the method, which it seems to me may be followed with success in the study of the greater part of the phenomenon of nature.

FOURTH PART.

ON STATISTICS.

LETTER XXXIV.

ON THE SCIENCES OF OBSERVATION, AND ON STATISTICS IN PARTICULAR.

THE sciences of observation have for their object the study of natural bodies, and of the laws connected with them. They proceed by the same paths, and the same rules are generally applicable to them.

They may be divided into three classes,—physical sciences, natural sciences, and moral and political sciences so far as their applications are considered. In all, it is equally found necessary to commence by collecting well-observed facts, grouping them with method and discernment, and weighing and appreciating their value. It is here that, properly speaking, the science begins.

The causes which might have produced these facts are next sought, their mode of action and their degree of energy studied. An attempt is made to proceed from a knowledge of what is, to a knowledge of what may be.

The physical sciences trace the general laws of the forces which act on all natural bodies, whether it be to change their peculiar composition, or not essentially to modify it. They in some measure abstract the individuality of each body.

It is not so with the natural sciences: they submit to an attentive examination the different bodies that are found on the surface of our globe, in order to class them and study their peculiarities. They are ramified in their development; and the ramifications ought to proportion themselves to the more or less complete organization of the individuals to be considered.

It is thus that Geology makes known the natural distribution of minerals on our globe, and Physical Geography that of vegetables and animals.

Mineralogy describes and classes minerals according to their characters, as Botany and Zoology describe and class vegetables and animals.

Chrystalography makes known the structure of inorganic bodies, —Anatomy that of organic bodies, from the smallest lichen up to man.

Natural sciences proceed then similarly; but a difference soon manifests itself. For organic bodies time is an important element, which causes them to undergo an uninterrupted series of modifications and transformations. They are born, increase, and die. Each phase of this development presents a separate picture; and the study of the laws which guide this succession form the object of a separate science, Physiology, to which there can be none analogous for inorganic bodies,—for these bodies do not live: they are, as it were, placed beyond the confines of time.

Man in his turn separates himself from vegetables and animals by peculiar faculties, which allow him to re-act upon himself, and to modify his moral and intellectual nature. He is eminently progressive; and science—that precious treasure, which belongs to him exclusively—permits him, beyond his individual life, to transmit to his descendants a mass of information, which is diversified according to time or place.

Plants and animals have remained the same as when they left the hands of their Creator. Some kinds, it is true, have disappeared, and others have shown themselves successively; but the description given of them at the Creation would have the same value now. It is not so with man, at least with the intellectual and moral man. History, especially that of the sciences and philosophy, retraces a series of phenomena which can but belong to him. The moral and political sciences are then exclusively in his domain.

Man is again distinguished by another privilege. He is eminently social: he voluntarily renounces a part of his individuality in order to become a fraction of a great body, which has also its life and its different phases. Such an aggregation of men forms a *people;* and when this people possesses a territory and a government, it constitutes a *state.*

States, like the individuals which compose them, are born, increase, and die. Their organization and their laws of development present a succession of phenomena which constitute their political history. The mean duration of states has not, as far as I am aware, ever been inquired into: it is true that it is very difficult to find their commencement and their end. The duration of a state is generally shorter than that of the existence of a people; in fact, the social compact may be rent, and the people, without being destroyed, be driven from the territory they occupy. The Jewish nation is a remarkable instance of this. The contrary could not take place; for if the people be destroyed, the state necessarily ceases.

If we consider a state during one of its phases of development, in taking its statistics, we in a manner arrest its march in order to study it more at ease, and to discover its organization and its relations with all which surrounds it.

Statistics then supposes a state to be for an instant stationary, so that the elements attached to its existence may be enumerated; whilst political history follows it in its march, and verifies all the phenomena it presents. The one science is to the other what, in a different order of things, statics are to dynamics, what rest is to motion. Generally, statistics relate to the present, leaving the past to history, and the future to politics. We should be wrong, however, in supposing this halt and this examination made by the statist as reduced to an infinitely short instant of time. The examination should, on the contrary, extend over a period long enough to eliminate accidental causes: we must, however, take care that it does not extend over so lengthened a period of time that the social state may have been sensibly changed during the interval.

To renew the comparison I just made: when a body is projected, we may seek to discover the line it has passed over; and if at one instant of its progress we could analyze all the forces to which it is subject, we could easily determine the tangent to the curve at such point,—that is to say, the direction it would take, but for any ulterior hindrances which might arrest it or cause it to deviate.

Statistics is not confined to a conscientious enumeration of the elements of a state so as to present (so to speak) its Anatomy : its investigations may with success be carried further, and may, as in Comparative Anatomy, enable us to make comparisons between the organizations of two states; or keeping within the limits of one country, observe a people at two different epochs of its existence, and compare the features they present, so as to discover what it has gained or lost, and the elements which have been most sensibly affected.

Comparative statistics does not then trespass on the domain of history nor of politics : it takes its place without the confines of time, and shows the position of two different people, or of one people at different periods of its history, in order to contrast its comparable elements.

If we take a higher ground than we have yet done, and lose sight of the peculiarities which characterize states, so as only to see the general bonds which connect them together, we should embrace the whole of the human race. We see in fact that states sacrifice a part of their individuality in order to constitute the great family—the general system of mankind—as individuals have done with regard to states.

The social body extends itself to the last limits to which man is able to penetrate ; and there is not a people, even in a savage state, which has not more or less developed notions of the rights of man. The progress of each becomes profitable to the others, and the benefits of science form a common treasure from which each may draw, and to which each contributes its share. The social body, like individuals and like states, has its peculiar life and its phases of development. General history, the history of science and that of philosophy, have the noble mission of representing it to us during successive ages in its threefold relation,— the physical, the intellectual, and the moral. But history is not

N

sufficient to enable us to perceive all: it transmits to us the knowledge of things accomplished, of phenomena of all kinds which have succeeded one another. Even *general statistics*, with the aid of Ethnography, only presents a more or less faithful picture of the social body at a particular instant of its existence.

This great body subsists in virtue of conservative principles, as does everything which has proceeded from the hands of the Almighty: it has also its Physiology, even as the least of organized beings. When we think that we have reached the highest point of the scale, we find laws as fixed as those which govern the heavenly bodies: we return to the phenomena of physics, where the freewill of man is entirely effaced, so that the work of the Creator may predominate without hindrance. The collection of these laws, which exist independently of time and of the caprices of man, form a separate science, which I have considered myself entitled to name *social physics*.

Physiology relates to individuals: it displays their laws of evolution during a determined epoch, that of their existence. When it refers to mankind it has no period, at least we know it not; and if there be one, we are so placed as to see but a small portion of it.

LETTER XXXV.

WHETHER STATISTICS BE AN ART OR A SCIENCE.

STATISTICS has then for its object that of presenting a faithful representation of a state at a determined epoch.

But, in reducing it to this form, should we consider it as an art or as a science ? Before replying, I will in my turn ask, Which was Botany at its birth,—an art, or a science? It was limited to the collection of plants, to their recognition, to their enumeration, and to the description of them in a more or less complete manner. How defective were the first inventories of the vegetable kingdom, even under the idea of an art ! and how little was it considered that Botany would one day become constituted as a science ! Who dreamed then of the ingenious and profound classifications since created? of Vegetable Anatomy, which should initiate us in the most intimate details of the structure of plants? and of Physiology, which reveals to us the mysterious phenomena of their development and their reproduction?

What, for their part, were Mineralogy, Zoology, and even Astronomy—a science so imposing in our days that no other can give a greater idea either of the creation or of the genius of man? However, Astronomy at its birth confined itself to taking an inventory of the heavens, to recognising the stars, to grouping them artificially into constellations, so as to impress them on the memory ; but how far yet were we, even after ages of study, from having measured the distances which separate the heavenly bodies, from having appreciated their magnitude, and even their weight, and from having torn away the veil which hid the laws of their motion through space, and of their preservation in the course of time !

But again, nearly all sciences in their origin, instead of producing salutary fruits, have given rise to most deplorable abuses. Astrology, by the aid of cheats and charlatans, boldly wrought on the credulity of men; while the true science of the stars, timid and unknown, was attempting its first steps, and was endeavouring to mount its usurped throne. Alchemy in turn seated itself by the side of the cradle of the science which studies the laws and composition of bodies; and for a long time it also deceived men by the promise of results which it had not in its power to rea-lize. Magic, next foreseeing the marvels which physics would one day produce, attempted to accomplish them in its own way, and equally to usurp a power which did not belong to it. Each science at its birth finds the same antagonism. So soon as we perceive the distant end to which it should conduct, so soon as we have the consciousness of its future, imagination seeks to seize by anticipation those treasures the contemplation of which should one day give us enjoyment: it gives birth to brilliant systems, and attempts to transfer to others the illusions by which itself has been seduced.

Such is the natural course of things. A science, before pre-senting itself in its true character, is compelled to undergo different phases: there are some which we have seen born, and whose births have been marked by the same seductions, by the same false promises. Denied by some, they have been exalted beyond measure by others: time only can bring them to their natural limits.

Statistics comes in its turn; and it arrives the last in the order of sciences, as man is presented the last in the order of creation: and scarcely has statistics given proof of its existence, before that is required of it which could not reasonably be asked of any of its fellows. Scarcely had it collected a few incomplete observations, before it was required to reveal the laws which regulate the pro-gress of the social body, and which assure its stability and its preservation. Professor Achenwald, in giving it a name, assigned to it a very modest mission; but statistics, in progressing, has shown that it is called to higher destinies.

I repeat, all sciences of observation, at their commencement, have undergone the same phases: they were arts, for they were confined to grouping, in a more or less successful manner, collections of facts belonging to a particular order of things ; and it is by the comparison and the study of these facts that they afterwards became elevated to the rank in which we see them shine in the present day. Why should we be more exacting with statistics ? If it still presents itself as an art to the eyes of the majority, its future is not doubtful to those who can consider the sciences of observation in a philosophical point of view. Its denomination, the fruit of an incomplete conception, has done injury to the idea which we ought to form of it, and has too much limited the circle of its attributes. But when an infant is born, and a name given to it, it is very difficult to say how it will be developed, and to know what future awaits it.

To consider statistics only as a point between two infinites, the past and the future,—to permit it only to describe minutely what it can have observed in so short an instant,—is evidently to restrict too much its mission.

The distinction is ingenious, but I doubt whether statists consent to recognise it. When, for example, we would compare the mean life of one state to that of another, or would compare it successively to what it has been among the same people at different specific periods, we should doubtless be still desirous calculating it, notwithstanding the opposing barrier, and historians would not think ill of it. Nor would the writers on the politics of states complain of the encroachments of the statist when, with actual data, he shall employ himself with the appreciation of the eventual state of certain institutions, and of the operations which they undergo.

A celebrated economist, J. B. Say, also restricted too much the limits of statistics, in saying that it was for this science to show the condition of things whose state might change successively, and not an immoveable state of things. If a country were composed of immoveable things, it would be sufficient to collect once for all detailed statistics ; but still it should be done. It would, besides, be difficult to prove the immutability of certain elements, if we were not in possession of statistics made at intervals sufficiently long to show the variations caused by time.

Collecting statistics is generally very well understood. But it is not so with the definitions of this science: there is almost always a tendency greatly to confine its domain. I think that the definition which I propose, and which moreover varies little from that given by many modern scientific men, sufficiently circumscribes the attributes of statistics, for it not to be confounded with the historical sciences or the other political and moral sciences which the nearest approach it. *It only considers a state during a determined period: it only collects the elements connected with the life of this state, applies itself to make them comparable, and compares them in the manner which is most advantageous with a view of showing all the facts they can reveal.* This study does not encroach on the domain of other sciences; and if this encroachment really existed, it could only be advantageous, under whatever denomination we might otherwise wish to hide it, that it might not trouble the symmetry of classification. Those who can really best understand the language of numbers (and it cannot be denied that under this form are summed the greater part of statistical data) are those who have collected them, who have examined them, and know their weak and strong points, and who are really accustomed to this kind of work.

To regard statists as machines charged to bring together rough stones, and to pile them indiscriminately on the place where the edifice is to be erected, leaving them to architects who do not know their value, and who most frequently do not know how to work them, is to expose ourselves to sad mistakes. There must be a unity in all things. Let the architect who constructs know how to collect his materials: let also the painter collect and arrange all that is necessary to make his picture.

LETTER XXXVI.

STATISTICAL OBJECTS.

THE general statistics of a state comprises essentially the five following divisions :—

1st. Population.

2nd. Territory.

3rd. The political condition.

4th. The agricultural, industrial, and commercial condition.

5th. The intellectual, moral, and religious condition.

Population is the statistical element *par excellence:* it necessarily rules all the others, since it relates, above all, to the people, and the appreciation of their welfare and their wants. It would be vain to attempt to form statistics of value, without taking as a basis the results of a census executed with all the care and precautions which so delicate an operation requires. The other data have no real value, except so far as they relate to the number of the population.

A census carefully made sums in a measure the solutions of the most important problems that can be proposed to a statist. The classification according to age allows of the establishment of tables of population, of forming correct ideas on mortality, on the forces at the disposal of the state in case of necessity, and of fixing the ratio between the useful fraction which contributes to the general well-being and the fraction which yet requires assistance and support to become in its turn useful. The classification by professions indicates the means by which the population provides for its subsistence, and tends to augment its prosperity : it allows the legislator more particularly to fix his attention on the principal wheels which work in the machine confided to his care. The classifications by civil condition, by origin, by education, furnishes the administration with no less precious information to assure internal good order, and to facilitate the execution of the laws.

Numerous difficulties are always connected with the operation of taking a census: we rarely find in administrations all the intelligence and zeal necessary to conduct and execute so important a work, and in the people sufficient understanding and a sufficiently complete absence of apprehensions and prejudices to give with exactness the information sought. If we consider that a census entails considerable expense, and occasions in a state a loss of time so much the more sensible from its affecting every individual at the same time, we shall see that this operation, otherwise so useful, should only be renewed at more or less distant periods. The decennial period seems, in many respects, the most favourable: it is at least that which is adopted in most countries.

A well-formed civil state is also one of the first wants of an enlightened people: it is indispensable to governments, as well as to the repose of families. To assure its regularity, it has been found necessary to pronounce punishments against those who, by negligence or ill-will, might introduce omissions or errors. The records of the civil state, in countries where such measures have been adopted, should be ranked among the most useful data which statistics employs. These documents in general relate to births, deaths, and marriages, and constitute what is generally termed the *movement of the population*. The registration of deaths according to age enables us to adjust tables of mortality, the advantages of which are felt not only in all the branches of the administration, but also in most questions relating to public health and to the operations of assurance societies. By comparing the number of births with that of deaths, and taking account of changes of residence, we obviate the necessity of frequent censuses, and are enabled to judge of the position of the population with all the exactness that can be desired.

When the movement of the civil state is established with care, it shows the ratio of legitimate to illegitimate births,—the number of stillborn, the respective ages in either sex at which marriage is contracted, the influence of professions, and a great mass of other information which interests the philosopher as much as the statesman.

The operations relative to the militia and the recruitment of armies are made, in some countries, with a regularity and a severity which should lead us to inquire as to the results they produce. These results are so much the more useful as they enable us to judge of the power of one of the most interesting sections of the population, that which is called upon hereafter to provide for its preservation and its defence.

I have already asserted elsewhere* that the numerical tables of a population, when they are made with care and with all the developments which science requires, are a fruitful source of instruction : they form, in the annals of a people, the most eloquent page that a statesman can read, if he understand them well. In fact it only belongs to the practised observer completely to understand the language of figures, and not to go beyond what they can teach him. Censuses well made, and which succeed one another on a uniform plan, at intervals sufficiently near, should present most precise notions of the physical and moral condition of a people, of the degree of its power, of its prosperity, and of the tendencies which may compromise its future : they would teach much better than voluminous inquiries, which are often fettered by prejudices and private interests, what we ought to think of the retrograde state, or the immoderate development of certain branches of industry.

In attributing so great an importance to the statistics of population, I am far from mistaking the value which should be attached to the possession of precise notions on territory and on political condition (which so powerfully influence the mode of existence and the future of a people), as well as on the creation and use of its riches, its moral and intellectual condition.

We may also see that, in treating each of these parts successively, we should abstain from collecting documents, interesting in other respects, but which are not directly connected with the object of statistics. The *territorial* part, for instance, ought not

* On the Census of Brussels, 1st volume of the *Annals de la Commission Centrale de Statistique.*

to comprise all the objects pertaining to the three natural king-
doms which the country contains, but only those used by man,
either for immediate consumption, or to be made useful by commerce and industry.

A particular plant may grow spontaneously in Belgium; but this
is no reason why it should have a place in the statistics of the
kingdom: if it have no reference to man, if it be neither useful
nor injurious to him, it should figure exclusively in the Flora of
Belgium. If hereafter man makes it useful, it is then only that it
will be inscribed in statistics; and the manner in which it will
be mentioned will be very different from that which the botanist
would employ. The statist will disregard its scientific character-
istics, and will only record the quantity that is cultivated and the
advantages derived therefrom.

It is the same with animals. They should only figure in statistics so far as they are useful or injurious; and when they appear
there, they should be presented in a very different light to that in
which they would be presented in the natural history of the country. If horses be mentioned, it will not be to consider their zoo-
logical description, but the number of these animals that man can
dispose of for his different uses, and the values they represent.
The division of territory, the nature and the description of land,
the extent of forests, the circulation of water, will be equally
studied by the statist and the geographer, but under very distinct
aspects.

We may say the same of Meteorology: the elements which the
physician studies with greater care are not those which attract
the attention of the statist. The latter desires to know above all
things how they may influence man, and contribute to his welfare.
The former studies Nature, whose laws he seeks, independent of
the idea of the benefit which we may derive from it.

This distinction is essential, because a great number of authors
have introduced into statistics other sciences which are foreign to
it, as Physical Geography, Mineralogy, Botany, Meteorology, &c.

Others, on the contrary, have wished to restrict it to the
presentation of purely numerical tables, without dreaming that
there is information which it would be impossible to express

in figures. For example, an exposition of the political condition belongs essentially to the statistics of a country; we do not, however, know how to express it in figures. The same may be said of information relative to the moral and intellectual condition. The simple recital of what has passed in a locality at a particular time sometimes better teaches the moral condition of a people than all the numerical tables possible.

The desire of only presenting numbers often causes us to omit essential information as to the manner in which these numbers have been collected, and the circumstances which might have had an influence in making them more or less exact or more or less complete. We thus omit a part of the data necessary to resolve the most simple statistical problem, and to deduce the useful consequences of facts the causes of which we would wish to appreciate.

The receipts and expenditure of a country, the state of its debts and everything that relates to the condition of its treasury, deserve greater attention from the facility with which we may be led to the commission of very grave errors. Some persons have compared countries with respect to the mean tax paid by each individual, losing sight of essential distinctions. Thus in one state such a sum figures in the budget of receipts, and in another it is omitted because it is paid as city dues. We must not, however, judge too exclusively of the amount of tax by the figures in the budget. Your Highness knows very well that you might much diminish the annual amount of the budget of your state by suppressing what is paid for public instruction, for the fine arts, for the amelioration of the ways of communication, &c. Would this be a benefit? certainly not; for the result would be that each one, to continue in the enjoyment of his actual advantages, must at a greater expense provide individually for the instruction of his children, and for the repairs of the roads in the neighbourhood of his home. It has been justly remarked that those are the most civilized countries who pay proportionally the most to the government.

The agricultural, industrial, and commercial statistics of a country, even narrowed within the closest limits, would form an immense work, if we sought to enter into all the details. But a judicious mind will appreciate, without trouble, the place which should be given to each thing according to its degree of importance.

If there are some elements which it may be useful to ascertain from year to year, there are many others which it will suffice to present at longer periods, especially those which are only subject to slight fluctuations. The products of agriculture, which most directly serve for man's consumption, should be ranked among the objects of which it would be important to determine the quantity and the price each year : it is the same with fuel, which forms so great a part in the wants of industry.

In the immense volumes of statistics published each year, by the different states, how little of the information is truly useful ! Innumerable subdivisions are made, and numbers so small obtained, that the consequences which may be deduced will be necessarily false : accidental causes are allowed to predominate to such an extent that it is impossible to separate them from the regular causes, when it is wished to appreciate their influence. This luxury of figures, this species of scientific quackery, occasions also considerable expense to the state.

One of the greatest inconveniences of industrial statistics is that it requires the intervention of persons who are almost always interested, or think they have an interest, in disguising the truth. When the government collects them, it is generally opposed by the manufacturer, who supposes it done with fiscal views. The desire to obtain favour for his industry, and to obtain what are called protecting laws (laws which are at bottom but real privileges obtained from neighbouring governments at the expense of other branches of industry), almost always tends to exaggeration in one direction or the other.

Governments also publish documents on importations and exportations. These tables, which are useful to consult, nevertheless often contain very vague returns : they are generally confined to the fixing either of prices from faulty valuations, or of quantities without considering either price or quality. In the official valuations, moreover, we only know a part of the truth : it is especially here that information, not susceptible of reduction to numbers, becomes necessary, in order to determine the probable quantity which escapes the legally stated values.

The statistics of the moral and intellectual condition of a people presents still greater difficulties; for the appreciations can only be founded on facts much more contestible than those given by industry and commerce. When we say that a province produces so many quarters of corn, so many gallons of oil, we know that the figures may be more or less in error, but we understand the nature of the unit. It is not the same when we say that a province produces annually so much crime: in addition to the existence of much uncertainty in fixing the numbers, crimes in general are not comparable in regard to gravity; and we often only perceive in a confused manner the relations which exist between them and the causes which have given them birth. But these relations it is well to study, if we wish to determine the moral condition of a people. Infinite precaution and sagacity are necessary to read with success the statistics of tribunals; for the documents they contain are very complex in their nature, and almost always incomplete.

What a mass of errors have we not accumulated in treating of pauperism! To probe this leprosy of society, we have had recourse to lists of the poor, and very often without inquiring if these lists were complete and comparable in different countries, or even within the limits of the same country. Real poverty is nearly always very different from the poverty officially returned. From a locality having no relieving-office, it is generally concluded that it contains no paupers; and according to similar estimations, founded on official lists, poverty will be in proportion to the sums distributed. Here especially we see the inconvenience of only consulting numerical tables, and of attributing the same value to all the units they contain. In Belgium a man will enter his name on the list of paupers to escape serving in the civic guard, or to obtain other advantages, without receiving a farthing of public benevolence. Can we then reasonably confound him with his neighbour who lives regularly by the alms he receives?

The greatest inconveniencies in moral and intellectual statistics relate then to the difficulty of rendering the units comparable. To measure the instruction of a people, or the state of its enlightenment, tables which show the number of children sent to school, or the number of persons who can read and write, according to the declarations made before recruiting councils or before tribunals, have been made use of. These documents are certainly very curious and very useful, but they only constitute a part of the information necessary to decide the question: we should further know what is taught in the schools, and what is meant by knowing how to read and write. This last talent is more frequently acquired in a mechanical manner, as the handling of a plane or a needle, and does not contribute more to the development of the intellect than to the formation of morals.

The evil here, I conceive, only exists in the bad interpretation given to statistical documents. But it is always the case that this bad interpretation may generally be referred to the incompleteness or bad arrangement of the table, or the numbers being presented without an explanation of the meaning to be attached to them. These difficulties should not arrest the statist: he will only feel the necessity of proceeding with the greatest reserve, of furnishing all the necessary means to render the documents he collects comparable; and if he cannot present them complete, he will at least try to indicate the probable limits between which the true numbers will be placed.

We should, in general, abstain from making use in statistics of data which are not perfectly exact; or if we are obliged to have recourse to numbers which do not present the necessary guarantees, we must at least take the precaution to provide for them.

LETTER XXXVII.

ON THE DIFFERENT FORMS ASSUMED BY STATISTICS.

STATISTICS are *general* or *special*, accordingly as they relate to all the component parts of a state, or some parts only, with a view of throwing light on certain questions.

When a statistical account only embraces in its particulars a small extent of country, it is called *local* statistics ; while the name of *universal* statistics is given to that which relates to the whole universe.

Statistics are also of two kinds, in reference to the sources from which they emanate. Some are published by government, or under its auspices,—others by private individuals.

Some writers have further distinguished the sources, according to the degree of confidence they merit, into *primary* and *secondary* sources.

Official statistics (that is, those published by governments) are in general special statistics, and emanate from one branch or other of the administration. In some states, and especially in constitutional states, these publications are periodical: thus we have publications respecting the finances, the movement of the civil state, the progress of justice, external and internal commerce, education, benevolent institutions, &c.

These publications should be free from every kind of discussion, and should be confined to the presentation of facts in their most simple form, abandoning them to the appreciation of learned men and statesmen.

I do not, however, think that rigour should be so far extended as absolutely to interdict all comparison of numbers. Thus the information respecting the course of justice which is published in France, in Belgium, and in the grand-duchy of Baden, generally gives, by way of introduction, in addition to the numbers for the particular year, those which have been obtained during preceding years, without however carrying the comparison further.

The relations between certain numbers which are most often compared together might also be established, in order to relieve those who would most often make use of the results from the necessity of repeating tedious calculations : thus, in the statistics in question, we should calculate the ratio between the number of condemned and the number accused or committed. In the statistics concerning the movement of the civil state, we should calculate the ratio of the number of deaths to the population in order to find the mortality, or the number of legitimate births to marriages so as to form an idea of the fecundity; but we must always be sober in such comparisons.

Governments, in publishing the documents collected, should always be careful to indicate the means which have been employed to obtain them, at least if these means are not already defined in laws and regulations still in force.

We very often see, especially in constitutional countries, special statistics formed to throw light on certain questions to be submitted to the legislature. The documents are even accompanied with remarks, which tend to render the conclusions that may be drawn appreciable. But when the government suggests the line of discussion, we find ourselves more or less prejudiced against it : there is a tendency to think it has a direct interest in making the conclusions drawn from the elements it presents to prevail.

Private individuals rarely publish statistical data without accompanying them with remarks, or without forming the comparative statistics : they generally have in view the study of a question in all its phases, or the carrying out of an idea with which they are prepossessed. The uncertainty as regards their aim makes us suspect the documents they present. They ought always to cite carefully the sources from which they draw, and present all the necessary guarantees to ensure public confidence. In these instances more particularly authority becomes necessary.

When statistics takes a practical character much art should be used, in order to derive advantage from and usefully to interrogate the numbers: much skill and a particular tact are also necessary for the appreciation of the information we obtain. If we are pre-possessed with a systematic idea, it generally happens that we adopt with eagerness, as favourable to such idea, results which present anomalies, not by virtue of a determined cause, but by virtue of accidental causes. These mistakes often occur without the knowledge of those who commit them. But we may generally remark that writers who desire that certain systems should pre-vail, in resorting to figures use very small numbers. In fact, by only taking a few observations, and selecting our numbers, we can by the effect of accidental causes defend all possible theories. By the aid of such numbers, whose truth in other respects cannot be contested, we set our consciences perfectly at rest, and demon-strate nearly anything we may wish. This it is which always inspires so great a distrust in respect to special statistics, and which has done the greatest injury to science with persons who only judge of things superficially.

There is one fault against which we cannot be too much on our guard: it is that tendency which exists, as well in private sta-tistics as in certain official statistics, to make an immoderate use of numbers. On opening these books we are frightened at the idea of the errors they must contain, especially if we think of the difficulty we find in combining a few simple data which are exact. Many learned men would translate everything into the language of figures; the most insignificant details of the adminis-tration, even those of private life, would become the subjects of so many private inquiries. They will soon seek, from the manner in which the thresholds of the churches are worn, whether they are entered with the left or right foot. Certainly, such researches directed by a man of genius may lead to very piquante results in respect to the constancy of our actions, even in the most insigni-ficant circumstances of life,—they may excite as much curiosity as a literary work intended to afford relief to the mind; but they must not take rank in the circle of labours with which statistics are occupied. If we descended to such details, our existence would no longer suffice for the study and discussion of the most important facts passing around us, and our habitations would be exclusively invaded by the voluminous publications which all these futile researches would require. In comparing the infinite multitude of social phenomena with the narrow bounds of our intelligence, and with the slowness of our labour, must we not say with Diderot, "What is our aim? the execution of a work which can never be accomplished, and which would be far beyond human intelligence were it completed! Are we not more ignorant than the first inhabitants of the plain of Sennaar? We know the infinite distance between earth and heaven, and we do not cease to raise the tower; but must we not presume that the time will come when our discouraged pride will abandon its work?"

LETTER XXXVIII.

ON THE MANNER OF COMBINING STATISTICAL DOCUMENTS.

It may be easily understood that the difficulty of obtaining exact documents forms one of the greatest obstacles to the progress of statistics. In other sciences private individuals may combine, with more or less trouble, the observations necessary for their labours; but it is not so with statistics. The greater part of the documents required can only be obtained by governments, which have often neither time nor disposition to ask for them; and when they have acquired them are led by their own interests to keep them secret, or to publish them only in part, and sometimes even to alter them.

Governments founded on principles of liberty are the most favourable for the kind of study which we are now considering, because they find in publicity valuable means of verification. If it is generally impossible for the private individual to collect documents relating to an entire state, the means of control are always at hand when the administration publishes and the press is free.

In the composition of statistical works many essential things must be considered. 1st. The question to be put. 2nd. The forms to be filled in. 3rd. The means of control. 4th. The most advantageous form of tables designed for publicity. Your Highness will doubtless excuse me, if I avoid entering here into all the details which a course of statistics would require; but perhaps you will be interested in learning some of the principal conditions to be regarded.

Before demanding statistical information, it is well to study with care the question on which this information is to throw light: we must seek to separate the causes which govern it, and especially those which exercise the greatest influence. Without this preparatory step, it will be impossible to draw up the programme of the inquiries that should be made; we should expose ourselves to the reception of incomplete documents or useless details.

The principal considerations which should guide an administration as to the questions to be asked are the following:—

1st. Only ask such information as is absolutely necessary, and such as you are sure to obtain.

2nd. Avoid demands which may excite distrust and wound local interests or personal susceptibility, as well as those whose utility will not be sufficiently appreciated.

3rd. Be precise and clear, in order that the inquiries may be everywhere understood in the same manner, and that the answers may be comparable. Adopt for this purpose uniform schedules which may be filled up uniformly.

4th. Collect the documents in such a way that verification may be possible.

Administrations should carefully avoid fatiguing by questions, and above all by questions which do not bear the stamp of immediate utility, because they expose themselves to the risk of receiving no answer, or of obtaining only faulty documents. Nothing brings more disrespect on power than to demand what must in the end be acknowledged to be impossible; for it thus loses the means of speaking with authority, and of commanding obedience.

On the other hand, so soon as people think they perceive in the questions proposed a financial aim or an inquisitorial curiosity, they conceive distrust, and make no scruple of giving inaccurate returns.

Simplicity and clearness of demand, together with uniformity in the forms to be filled up, are essential conditions to obtain comparable results: without them no statistics are possible. When the question relates to ages, professions, diseases, it is of the greatest importance only to employ classifications perfectly identical, in order that the general information may be compared even to the slightest detail. The most perfect unity should reign through the whole.

It is to establish a like unity that, in certain states, such as Belgium and Piedmont, central commissions have been formed to collect and arrange the different elements which should be included in the national statistics. The necessity of such institutions is particularly shown where we see, in very enlightened countries, the principal departments sometimes publish very different numbers to express the same things, or make classifications which render comparison impossible. The least inconvenience of this want of uniformity is the double labour which, while useless expense is occasioned, augments in an inconvenient manner the volumes of official publications.

It is well also to distinguish carefully, as I have already said, the elements for which the statistical data must be annually combined. I rank first the elements which most relate to the prosperity of the people,—those especially which are exposed to the greatest fluctuations: it is sufficient to furnish the others at more distant intervals. The census of the population, notwithstanding its importance, is one of this number: the difficulties which surround operations so delicate, and the consequent expense, must necessarily render it less frequent.

We should also endeavour to determine the number of observations necessary to establish a fact. Thus, to exhibit the salutary effect of vaccination, a much smaller number of observations is requisite than to discover the preponderance of males in the number of births. Those who are charged with the collection of statistical documents should carefully study all these gradations, so as not unnecessarily to multiply the labours of research, by admitting insufficient or completely useless details. Economy of time is, in fact, a principal point in administrative cases; and in many circumstances we should prefer it to economy of money.

There are some elements, very useful to be known, which can only be indirectly determined. It would be imprudent, for instance, to inquire of the cultivator the net produce of his land: this would excite distrust, and provoke incorrect answers; while the same cultivator would feel less difficulty in stating the nature and quantity of his produce. Laplace proposed to substitute for the general census of a great country, such as France, some individual censuses in chosen departments, where this kind of operation would have the greatest chance of success, with a view carefully to determine the ratio of population to births or deaths.   By means of these ratios, and the returns of births and deaths in all the other departments—returns which could be given with sufficient exactness,—it would become easy to determine the population of the whole kingdom.   This mode of operation is very expeditious; but it supposes an invariable ratio in all the departments. We may, it is true, so select the departments, in respect of which we determine the ratios, that the mean shall not vary much from the ratio we should obtain for the whole kingdom.   This indirect method should be avoided as much as possible, although it may be useful in certain cases where the administration has to proceed with rapidity.   It may also be used with advantage as a test of accuracy.

Not to secure facility for the verification of the documents we collect is to miss one of the principal aims of the science.   Statistics are only of value according to their exactness: without this essential quality, they become useless, and even dangerous, since they conduce to error.   The necessity of checking statistical documents is so great that I propose making it the object of a special letter.

Before I lay down my pen, it yet remains for me to offer some remarks on the most advantageous form for statistical tables intended for publication; and I will first say a few words as to their arrangement.   This question may appear very futile, especially at a time when we carry in our books all the anomalies which we affect to show in our garments.   "What matters," say they, "whether it be quarto, or modest octavo, or the majestic folio?" The choice to be made is not so futile as it seems at first glance. You shall judge.

The statist has especial occasion to possess the means of making comparisons, and of having the greatest possible number of facts under notice.   It is important that numerical documents should not be too much divided by the partial additions we are forced to make at the bottom of each page.   The largest tables then would be the most advantageous, if we could handle them conveniently, and at one glance seize upon all they contain.   It has been attempted to combine all these advantages, by publishing on an extremely reduced scale, and by using very small figures; but a new inconvenience was presented, which was greater than that which it was wished to avoid. The figures were illegible on account of their smallness; and without the use of glasses it was impossible to read them. Experience seems to have proved that the quarto form best suits all exigencies,—the field of sight more easily embraces the whole page, which may moreover contain, without confusion and in very legible characters, a great number of documents which it would be desirable not to separate.

Did I not fear to descend to too minute details, I would also speak of the disposition to be given to tables, and of the regulation of the headings, which above all require precision and clearness.

Some people, in calculating arithmetical means, carry the operation much beyond the limits of exactness, and give decimals to which no confidence can be attached. Thus an edifice may have been measured six times in succession, and its height may have been found to be about 60 yards. We will suppose the six measurements to be expressed in yards and feet. A calculator afterwards takes the arithmetical mean, and expresses it in yards, in feet, in tenths, hundredths, and thousandths parts of feet. This is evidently showing too scrupulous a conscience, for such a mean can inspire no confidence in its last decimals. How can we reckon on the exactness of a thousandth part of a foot, when we have only six measurements, each incorrect by half a foot? These faults are frequent in works which treat of the sciences of observation, and in statistics in particular.

It must be admitted as a principle that the decimal should not be carried beyond the figure where doubt begins. Physicians will give specific gravities with a number of decimals evidently greater than the sensibility of balances would justify. What must we conclude? Must we charge them with quackery or ignorance? It is the same case as asking, with a celebrated chemist who saw such an exaggeration of exactness committed, "What do you complain of? It may be that these last decimal figures which you contest are very exact, and that the first ones only are faulty."

Often this great number of decimals, which very evidently

exceeds the limits of exactness, is only given from a certain habit, and without the possibility of good reasons being alleged for their use : we could cite examples in many works on Physics and Astronomy, even in those which have done the greatest honour to our age. Numbers are given as they have been calculated by means of tables of logarithms. It would, however, be well to suppress all the figures which are useless, and which might induce an error as to the degree of precision.

The Theory of Probabilities indicates the limits at which we ought to stop in such a case. Perhaps it will be more convenient to recur to this so simple principle, deduced also from theory, that precision increases according to the square root of the number of trials. Thus, when you have weighed an object many times in succession, and each weight is doubtful in the first decimal, the individual weights should only be given with one decimal,—the mean of them has two, three, or more decimals, according as the number of observations might be under 100, or from 100 to 10,000, &c. The degrees of precision are in general as the roots 10, 100, &c., in respect to each weight taken individually.

There is another error very generally made in works on statistics, which it is well to notice in passing. If I wished to know the mortality of the department of the Seine, I should compare the number of deaths in a year to the number of the population; and the ratio would give me the mortality of the department. I could make the same comparison for each of the other departments of France, and should thus obtain 86 ratios. If, however, I wished to form a general idea of mortality in France, should the mean of the 86 ratios calculated in this manner be taken? Certainly not. All the deaths in the kingdom must be compared to the general number of the population,—to do otherwise would be to give the same importance to each of the departments, whatever their population might be. The mean calculated in this way may differ very sensibly from that which would be obtained by the other process of calculation.

There is, however, a distinction to be made, which may appear subtle at first, but not the less founded. In a particular case the mean of the ratios must be taken as the mortality. This case is when we wish to form a general idea of the mortality with reference to the different regions of France; in which case it is less the number of the population that we must consider than the extent of the departments. Each ratio given by a department assumes the same importance, although it may be based on a greater or less number of observations.

LETTER XXXIX.

MOST men accept with equal confidence all statistical documents. Whatever the source whence they proceed, the mode employed in their collection, the number and value of the observations, they attach to them the same degree of importance. I have often heard statistical results quoted by men of the world; but I do not think I ever heard it asked how far they were exact,—whether they rested on observations sufficiently good and sufficiently numerous to be admitted without restriction. The same levity generally exists among writers. It can, however, be easily understood that this want of discernment and criticism cannot lead to sound conclusions. To make an edifice solid and durable, the architect should examine with care the materials he employs in its construction.

Statistical documents involve two kinds of examination,—a moral examination and a material one. The first can be made independently of information received: it refers, in fact, to an inquiry as to the influence under which the information has been collected, and to the appreciation of the value of the sources from which it is derived.

During the war of independence, the United States carefully misrepresented the true number of their population : they exaggerated considerably the number of inhabitants in maritime cities, in order to put the enemy on the wrong scent. Assuredly, no good appreciation of the American population could be founded on the documents of this period.

We should also take care not to value the importance of the commerce of lace between Belgium and France by the import tables published annually by the Customs. The high duty, and the facilities for fraud, must produce in these tables great omissions. Good sense is sufficient to reject, or only to receive with extreme reserve, numbers which have necessarily been altered either by general or particular interests.

We should generally place more confidence in statistical data which can be obtained easily, and without injuring any interest,— without wounding any susceptibility in those whose utility and convenience every one can appreciate, especially in those which interest the repose of families, and which are collected in virtue of the laws. Under these different relations, the tables of the civil state in Belgium and in France should be placed in the first rank, when the local authority discharges its duty properly.

Ignorance of the customs and laws of countries sometimes leads to singular errors. Thus, by means of official numbers, M. Sarauw pretended to prove that in the island of St. Croix, in the Danish Antillas, the mortality of the black slaves was less than that of white men even in Europe; and this assertion might appear so much the more imposing as M. Sarauw resided in the island in question. However, Professor David of Copenhagen has proved that this opinion, which was held in good faith although contrary to anything before observed, rested only on a mistake, or rather an omission. In the Danish colonies, in fact, there is a poll-tax on negroes; but an order of the 16th July, 1778, exempts negro children which die before attaining one year of age from this tax. The result has been that children under one year have not been registered in the lists of births or deaths of the black population; and as the mortality is considerable during the first year of life, the black population has appeared to have a marked advantage over the white.

Before deciding on births and deaths of a country, it is then very important to know how the documents are collected, and what guarantee there is for the regularity of the registrations. In many countries the births are taken from the parish registers, which do not include Jewish children. In some states, again, the stillborn are comprised in the lists of births,—in others they are separated. By the neglect of all these circumstances the most strange results are constantly obtained.

To whatever they may refer, the documents of the movement of the civil state are generally the most complete of the statistics of civilized countries. The information referring to an enumeration perhaps deserves less confidence, especially that relating to houses, because the officers by whom they are furnished almost always suppose the authorities to have fiscal intentions.

On the other hand, if private individuals are on their guard against the government, the latter also is reserved towards them, and often only communicates to them a part (more or less complete) of the information collected. Statistical data may then be altered in their source, and in the channels which transmit them to us. I would not employ myself on the verification of the figures of a census, if I knew that such census was undertaken with the avowed end of raising troops, because I am persuaded that a double cause of error would pervade the whole operation. I would also avoid the use of returns of births and deaths, in a country in which it should be proved to me that such returns were not made according to a regular plan nor with any guarantee.

Very often, notwithstanding the zeal of individuals and the cares of government, statistical tables remain necessarily incomplete: such are those of the tribunals; such also are those of the Customs and Excise. It is difficult then to establish from these tables a good moral appreciation, which ought however to precede every other check.

The material examination of statistical documents does not require less prudence and sagacity; but here at least we find some rules to guide us, especially when the information may be translated into numbers,—information of which statistics make most frequent use.

It is, above all, necessary that the numbers compared should be sufficiently large, in order that we may believe them to be uninfluenced by accidental causes. All results do not require the same number of observations to present the same degree of exactness: thus, as I have already observed, many more observations are necessary to show the preponderance of one sex in the births than to establish the efficacy of vaccination against small-pox, or the influence of age on deaths.

However great the number of observations, it becomes insufficient when there is reason to believe that periodical causes, or a very preponderating accidental cause, may have exercised an influence. If we desired to know, for instance, the mortality of France, although the number of deaths in a year may be considerable, it would be imprudent to be content with that, because maladies might fortuitously increase the value. A year may be as fatal to men as to the fruits of the earth: the results of two years at least should be employed. However, this same number of deaths when it refers to a town, or even to a department, would be more than sufficient, because it would be the result of many years in which the accidental causes might have compensated one another.

Generally, to ascertain, without recourse to mathematical theories, the degree of precision attained in the calculation of a mean, it is sufficient to divide the whole of the values observed into two or more groups, and take their means separately; and if these means differ but little one from the other, they may be regarded as precise.

I have already sufficiently explained the use of the tables of precision when we wish to determine the exactness of a mean: I think I need not again speak of it. I will only remark that it is well to verify, at the same time with the mean, the extremes between which it is comprised.

Material verifications should extend to the figures themselves: it is even prudent to examine the additions before entering into any discussion. I beg your Highness' pardon for entering into such minute details; but there is often occasion to regret the omission of these little preliminary precautions, and we are forced to revise results which were believed to be exact. When a column contains many figures, it would be advantageous after the general addition to perform two partial additions, the sums of which should agree with the general addition. Sometimes the addition of vertical columns are verified by those of horizontal columns.

After these different verifications, we must examine carefully the continuity of the numbers which express the same thing; seek if they present any sudden changes; and should we find it so, appreciate whether they are owing to errors in the figures or to accidental causes. Thus, when I study the mortality of a town, I rapidly examine the series of the numbers of annual deaths of which the period with which I am engaged is composed. If I find a number too small or too large, I check the calculations : if they are right, I check the sources from whence the original numbers were taken ; if they also are exact, I seek whether the variation remarked is due to accidental causes : or whether it depends on constant ones.

This study is singularly facilitated by diagrams. A simple line allows us to appreciate at a glance a succession of numbers which the most subtle mind would find it difficult to retain and compare. The facility we thus have of seizing upon a series of results, and of recognizing their progress, is such that we may at the same time follow a series of results of another order, and judge by the common inflections of the lines the ratios existing between the elements under comparison, and the common causes which may modify them simultaneously.

Diagrams not only afford relief to the mind,—they also give to the study of phenomena the same advantage that Algebra has introduced into calculations : they generalize and allow of abstraction.

If, for instance, we wished to know the influence which the price of grain might exercise on the movement of the population, we should construct four lines which, by their undulations, would express the fluctuations observed from year to year in the price of grain, in the numbers of births, deaths, and marriages. We should generally find that when the price of grain was high the number of deaths increased, and those of births and marriages decreased : so that the first two lines present a certain parallelism; and the same with the others, but in a different direction to the first. As however the price of grain is not here the sole influencing element, and as, moreover, births, deaths, and marriages are not equally affected by the same causes, the inflections of the lines do not always correspond at the same period. We have then the valuable advantage of perceiving at first glance the least anomalies, and of being able easily to seek for their causes, whether these anomalies proceed from errors, or whether they ought to be attributed to real causes which have left their traces on the social state.

We must not, moreover, lose sight of influencing causes which do not produce their results immediately. We have found, for example, that the traces of years of scarcity or of abundance are not shown in the movement of the population until about a year afterwards. Sometimes even active causes do not produce their effects until a still later period.

Notwithstanding the objections that have been made to graphical methods by some writers, who have perhaps been too much struck by the abuses to which they have given rise, I think they cannot be sufficiently recommended, when it is wished to check series of numbers influenced by common causes, and to take a general view of the modes of action of such causes.

LETTER XL.

When we have carefully proved that a series of documents are morally deserving of confidence, and the figures have been submitted to a rigorous verification, it remains to make use of them.

It would be absurd, in fact, laboriously to collect statistical data for the sole purpose of swelling out volumes. We may then confine ourselves simply to the verification of results deduced from the documents compared, or to the interrogation of the numbers with the view of enlightening a particular question.

In the former case the numerical documents will be grouped in the different arrangements of which they admit, and in a manner to show the causes by which they may be modified. In making these arrangements, we must guard against any preconceived idea of the nature of the final result. It is useless to add that we should assure ourselves whether this final result rests on a sufficient number of observations for us to rely on its value. If we have taken all these precautions, we have only definitively established a fact which we must know how to interpret.

This is doubtless much; but it yet remains to explain the fact, and to remount to a knowledge of the causes which have produced it. This search may offer great difficulties, especially if we propose to appreciate the degree of energy of the causes. To make myself better understood, I will have recourse to an example: I suppose that I merely possess the result of a good census of a country, and I will try to indicate some of the conclusions that can be deduced.

I admit that a careful examination of the documents of the census, both in a moral and in a material respect, has been already made, and that there is thorough reason to be satisfied in this double point of view.

The first idea which will present itself will be to separate the total number of the population into two parts, indicating the number of men and the number of women. These two numbers generally differ but little in countries of a certain extent. After having established the general ratio, we must seek what it becomes in the different parts of the country: thus in Belgium we should have to calculate its value in each of the nine provinces.

If the ratio is sensibly the same, we should have reason to think this census homogeneous, and the result observed deserving of confidence; for the probability that these nine ratios should be sensibly the same in a purely accidental manner is extremely small.

If the ratio of one province differs from that of another, we must seek the cause. The principal one may be the nature of the census, the smallness of the numbers compared, a purely local cause, &c. The preliminary examination of the statistical documents should enlighten us as to the first presumed cause of error. For the others the numbers may be divided again, and a search made as to where and how the ratio is modified in passing from one province into the other. If doubts remain after this research, the rest of the investigation might aid in clearing them.

A great numerical difference of the sexes, in passing from one province to another, shows in the population a want of homogeneity which ought necessarily manifest itself under other aspects: thus the census will show the classification of the population by ages in each province, distinguished by the sexes. This new classification shows us, in the case of a numerical excess of one sex over the other, at what ages this is more particularly remarkable.

Industry and commerce may alter the ordinary proportions, and call a greater number of men to certain localities. Thus at St. Petersburg the number of men considerably surpasses the number of women; but this circumstance is owing to their domestic condition, which is not the same as with us.

The classification by ages not only offers the means of verifying the details of the census with respect to the sexes,—it affords most valuable information on the intrinsic value of the population, and allows us to form just ideas in the estimation of the useful part of which it is composed. Two countries may contain 4,000,000 inhabitants each, and be in very different positions as to the value of these populations: we must therefore find the useful predominating element,—that is, the comparative number of adults.

I can, without preconceived ideas, divide the population of each province into three great groups,—children, persons of full age, and old men,—not to descend to too minute categories. If these three classes of individuals are evidently in the same ratio throughout, this is a new proof of the homogeneity of the population, and the goodness of the census. If they present palpable differences, the causes must be sought.

In examining a census, we shall doubtless not lose sight of the influence of living in towns, or in the country, on ages and sexes. The passage from one province to another presents this advantage—that the general number is divided into many series, which may serve as a mutual check.

It will be the same for the civil state. The relative numbers which indicate the single, married, and widowed, for each sex, should be examined with the greatest care.

In a kingdom where the inhabitants are throughout in identically the same circumstances, we might say *à priori* that all the categories which I have just designated should be proportionally the same, except the little inequalities introduced by accidental causes; and reciprocally, if the categories are the same, and the numbers large enough for us to suppose accidental causes to have little influence, we may conclude that the inhabitants live under the same conditions. If the elements are essentially different, it is because they are influenced by local causes; and these causes are so much the more energetic as the elements are more dissimilar. The study of the most modified elements, in general, furnishes the means of discovering active causes and their degree of energy.

This example may furnish a first idea of the manner of discussing a series of observations, and of discovering the different consequences to which it conducts. I will not insist on all the other combinations that may be made of a well-arranged census.

I will now suppose that I interrogate the numbers of the census with a determined end, and with a view to enlighten a particular question. This kind of discussion is more difficult, and requires more reserve, than that with which we have just been engaged. I will admit, for example, that I only use the numbers of a census to ascertain how far industry may be of advantage to a country. It may be seen that the question here is the physical condition of the population.

If all the country were equally industrious the question would be nearly insoluble, since of the two elements of comparison one would be wanting. We might, however, if the census gave the professions, form two or three large groups,—for instance, the manufacturing classes, the agricultural classes, and the remaining inhabitants. These groups would contain the individuals of each category, with their wives and children. Then we should establish between them the same comparisons as those previously indicated, whether it be to show the differences of the sexes, of ages, of civil condition, or again to discover the proportional number of those who live at the expense of public charity, as was done in the census of the population of Brussels in 1843.

More generally a country presents a manufacturing and an agricultural population. It is then between these two fractions that the comparison should be established; but foreign causes, which might mingle their influence with that of the causes which we wish to determine, should be thrown out.

There are some stumbling-blocks to be avoided, on which many statists (too little on their guard against the digressions of their imaginations) have fallen. The principal are the following:—

1st. Having preconceived ideas of the final result.

2nd. Neglecting the numbers which contradict the result which they wish to obtain.

3rd. Incompletely enumerating causes, and only attributing to one cause what belongs to a concourse of many.

4th. Comparing elements which are not comparable.

I will speak more in detail of each of these causes of error. This task will perhaps be very difficult to accomplish, especially if I revert to myself; for I cannot guarantee that you may not find in my own works examples of what I think right to blame. However, I must not here engage myself with personal matters. The question is, in fact, much less whether I may have formerly been wrong than whether I am now right.

LETTER XLI.

SCIENCES at their birth undergo the same phases: they please by their novelty: every one thinks himself called upon to give them his patronage, and to associate himself with their first successes. They have especially a powerful attraction to men of the world, who are charmed with giving themselves scientific airs, without having to make too great an effort of mind, and without being obliged to grow pale over books. A new science, moreover, has no annals: it supposes few facts to be studied, few works to be read: all that it can teach is soon learnt. Thus geology, political economy, statistics, have successively had many proselytes, more especially among dreamers of all sorts who have wished to make use of them in support of their visionary ideas.

In fact, many persons only recur to statistics with a view to give consistency to preconceived ideas: they wrap up their ideas in a scientific veil in the hope of rendering them imposing. We see them employ, without examination, all the documents which seem favourable to their views, and reject the others as little worthy of confidence. In acting thus, some may do so in good faith, and only sin through ignorance.

With such eclecticism we may find in statistics means of defending almost every position. It is these abuses of science which have given rise to doubts as to its utility. If in medicine some new system arise, almost at the same time clinical results appear in order to prove its excellence or its nullity. On the one hand all cures are carefully cited, and all failures are attributed to other causes more or less specious; while on the other hand only the unfavourable cases seem to be borne in mind.

In politics especially statistics become a formidable arsenal from which the belligerent parties may alike take their arms. These arms may be accommodated to all the systems of attack and defence. Some figures, thrown with assurance into an argument, have sometimes served as a rampart against the most solid reasoning; but when they have been closely examined their weakness and nullity have been discovered. Those who allowed themselves to be frightened by such phantoms, instead of looking to themselves, prefer rather to accuse the science than to confess their blind credulity, or their inability to combat the perfidious arms that were opposed to them.

During the most disastrous combats of the empire, it was undertaken to demonstrate, by means of figures, that war is favourable to the development of population, and that the French people were never in a more prosperous condition. However, the flower of the nation was annually decimated; and those who escaped danger, borne down with labour and fatigue, ended in their homes a premature old age. In glancing over the lists of the population, we find even now the gaps which war has left there.

As for the rest,—that worthless documents are collected, and that an improper use is made of them, are inconveniences from which other sciences are not more exempt than statistics. The examples are less numerous only, because the route is already better traced, and because there are more points to which to repair in order to discover if we are in error.

In statistical documents no number should be neglected which is connected with the particular question,—at least unless we have reason to doubt its value, or unless it be too small to be usefully employed. Even in this case the motives for abandoning it should be stated.

We see persons profoundly convinced of a truth seek to establish it directly by the authority of figures, and give as they think a mathematical demonstration. However, by means of the statistical documents which they unskilfully employ, they most frequently produce an opposite effect to that which they desired. Thus we cannot reasonably doubt that enlightenment contributes to man's happiness, by illuminating his intellect and fortifying his morals. In the attempt to demonstrate this, what has been done? It has been thought necessary to establish that the number of crimes is inversely as the number of children sent to school,—as if the number of crimes, even were it known, had as its only cause the greater or less development of the intellect, and as if the development of intellect was measured by the number of children sent to school. What has been the result of this? It has been found, after well examining statistical documents, that the number of crimes is more generally in a *direct* proportion to the number of children sent to school than in the *inverse* proportion. The conclusion is exactly the opposite of what was at first desired, —a new error, which some have with the same levity admitted.

I have already had occasion to observe that the phenomena which relate to man are very complex: they are modified under the influence of an infinity of different causes which must be ascertained and studied with care. Most of the accredited errors of statists proceed from an incomplete enumeration of causes, and of their inability to attribute to each the degree of importance which is due to it.

The morality of a people is not a thing that can be directly appreciated: it can only be judged of by effects. When we see a bar of steel, it is impossible to say whether it is a magnet, and especially to fix its degree of energy,—we must necessarily have recourse to experiment. But morality is exhibited by good or bad actions,—and hitherto, to estimate the morality of a people, bad actions only have generally been examined; again, of these such only have been taken into consideration as have been punished by law as crimes or offences. In my next letter I propose to show how much this latter information leaves to be desired, and how incomplete it is even in countries where it is collected with the greatest care. I will only here consider the causes which give rise to crimes.

Man has at his birth all the tendencies which may cause him to deviate from the line of duty. These tendencies develop themselves, or are modified by the influences amidst which he is thrown. First, the early education, its religious principles, the example of parents, the degree of comfort, the wants which he later feels, his relations,

the state of his intellect, his profession, the customs and the legislation of the place where he dwells, should be considered as so
many causes which either draw him to or drive him from the
abyss in which he may be lost. When we are considering one
person in particular, it would be impossible to state *à priori* what
cause exercised the most powerful influence to precipitate him
towards evil: it is only after having seen him yield, and after
studying his previous conduct, that we can discover the cause
which has more particularly led him astray. If such a study, pursued with care and without prejudice, lead to the conclusion that
each crime had always the same cause,—want of instruction for
instance,—we must conclude by admitting that it is this cause that
determines crime, and that the other presumed causes are without
influence; but it is not so. We see successively an infinity of
causes prevail, with in truth very different degrees of frequency
and energy.

In studying the crimes of a country, the question then is not to
find one single influencing cause which predominates over and
effaces all others,—we must learn how to embrace the problem in
all its generality. The state of instruction is certainly a very
influential element; but it cannot be isolated without imprudence,
because its influence is most frequently effaced by that of other
more active causes. There are even crimes which suppose, as essential conditions in those who commit them, a very high degree of
enlightenment.

Those then are wrong who have exclusively considered the
absence of instruction to be the general cause of crimes, and especially in having taken as a measure of information the number of
children sent to school without even considering what they learned.
The contrary position is still more absurd,—because in some provinces or in some towns the number of crimes has been found to
be directly proportional to the number of children sent to school,
it has been endeavoured to insinuate that the development of information was contrary to the development of morality. No regard
has been had to the fact that in these localities the greater number
of crimes was generally owing to the greater accumulation of
inhabitants and wealth, which facilitated crimes against the person

and against property at the same time that it explained the greater attendance at schools.

We must also remark that the number of crimes may depend as much on the degree of morality of the inhabitants as on the legislature of a country. The reform of certain laws has sometimes sufficed to produce the suppression of an entire class of crimes, or at least to cause considerable reductions in their number.

Conclusions are too hastily arrived at; and in wishing to make statistics complacently apply to the demonstration of problems of which it had not the elements of solution, and which it was not called upon to resolve, considerable wrong has been done to the science. Even most judicious minds do not always distinguish the abuse they have made of a science from its inability to solve certain difficulties.

LETTER XLII.

I HAVE stated that no number must be neglected which is connected with any particular question. But a very strong objection is presented against most statistical documents, and against the results that may be deduced from them, from the documents being in most cases incomplete.

The number of the population is, without doubt, the most important statistical element. However, I do not think that there exists a single country in the world in which this element is well known. I do not only speak of a mathematical precision, but of such an exactness as can be admitted in the sciences of observation.

Each science, in fact, allows of a certain exactness, which may be discovered by the degree of approximation which it attains by its measurements. Astronomy, which is considered to operate with the greatest precision, may determine the position of the stars without having to fear error of more than two-tenths of a second in arc. But such an error in relation to a circumference, or in relation to 360 degrees taken as unity, falls to the fifth or sixth decimal.

By means of a good balance, constructed to weigh about a pound, we appreciate differences of the hundredth part of a grain; and the sensibility of the instrument thus extends to the sixteenth decimal, taking the body weighed as unity.

The specific gravity of a body is only obtained with an exactness which reaches the fourth or fifth decimal: it is nearly the same with linear dilatations of bodies.

The magnetic intensity of the earth can only be measured exactly to the third or fourth decimal. We may in general admit that the different elements whose magnitudes we appreciate in the sciences of observation are only determined with precision to the fourth or fifth decimal.

To proceed with the same exactness, it would be necessary that a census should not be in error more than one or two individuals out of 10,000; or for a population of four or five millions of inhabitants, as that of Belgium, that the error should not exceed 1,000 persons, or for France 8,000. But we are still far from attaining this precision; and it is my opinion that the population of Belgium is estimated at about one-tenth below its real value. This doubt then would influence the second, and perhaps the first decimal. The births, deaths, and marriages are better stated; and I think their number nearly attains the desirable degree of exactness.

In this state of things, should we reject all the results into which the number of population enters, although evidently faulty? I think not, especially when we would consider relative quantities rather than absolute. In Belgium, for example, when a subdivision for fixing the quotas of militia is involved, and the number of militia in proportion to the supposed population in each province is sought. But if we have reason to believe the supposed population is throughout inferior, in the same ratio, to the real population, it is evident that the subdivision will be as well made as if the number of inhabitants were exactly known. It would be the same with fixing taxes, which also might have the population as a basis.

There is another very striking example where incomplete documents may nevertheless be useful. I speak of those pertaining to criminal statistics.

What, in fact, are the principal documents which we possess to judge of the morality of a people? The accounts rendered by justice only. These accounts again only show the crimes prosecuted before the tribunals. But the crimes of a country may be divided into three principal categories,—those which are known as well as their authors; those which are known, but their authors are unknown; and those which are totally unknown to justice. Of all these crimes, we are only informed as to those which belong to the first category. We may then ask if it is possible to make use of such incomplete documents.

I admit without discussion that the crimes of a country, were they all known, would suffice to resolve the question under discussion. I accept this hypothesis to simplify matters. I even admit that all crimes have the same gravity.

But what advantage can we derive from the knowledge of those crimes only which have been brought before the tribunals? None, I think,—at least if this ascertained portion does not always remain the same in proportion to the total number. In Belgium there are annually three or four hundred crimes tried at the assizes; but this is perhaps only a tenth of the crimes annually committed. Now, if I were well convinced that this number of crimes brought to justice remained always the tenth of all the crimes committed, I could judge each year whether the number of crimes increased or diminished. We are then led to admit that there exists a constant ratio which gives the measure of the *activity of justice*,—a ratio which we must take into consideration, if we wish to compare Belgium with itself at different periods. Should we wish to compare it with another country, we must also be informed as to the activity of justice in such other country.

In the first place, is the ratio constant? I do not hesitate to answer in the affirmative; at least we may regard it as such if judiciary prosecutions are always made with the same activity—if the statistical registrations of the facts are made with the same exactness—if reforms in the laws do not change the penalties, and do not tend to correct certain crimes,—if, in fine, the state of the country does not undergo essential modifications: the causes being the same, the effects will remain the same also. The regular reproduction of the same facts shows itself in what is known, as well as in what is unknown. I had thrown out these ideas before they were demonstrated (in part at least) by experience.

Belgium prior to 1830 only gave the official number of crimes known and prosecuted: it has since published, for the seven years from 1833 to 1839, the number of crimes known, but which were not prosecuted because the authors were unknown. Now this latter number has proceeded from year to year with even more regularity than that of crimes prosecuted—no doubt the third category would present the same regularity, should a knowledge of it ever be attained.

During the seven years of which I have just been speaking, 140 crimes against the person have been annually tried before the tribunals, and 64 have remained untried because the authors were not known. This number is to the first as 1 to 2 nearly. The mean number of offences against property known and prosecuted has been 276, and that of offences known and not prosecuted 674. This last number is here the larger: it is nearly treble the other. Crimes against the person, being more grave, are those of which the authors can be best discovered: it is not then surprising to see so great a difference in the ratios. We may in general admit that crimes have so much less chance of being discovered and prosecuted as they are less grave.

It is also seen that, taking them without distinction, out of 1,154 crimes annually known to justice, 416 only, or little more than one-third, appear in the records of justice.

I think I do not exaggerate, in stating that the number of crimes committed of which the law is cognizant is not greater than that of crimes which remain wholly unknown. According to this estimate we find in the official returns only about one-sixth of the crimes committed in Belgium. But it is with this so defective statistical element that we presume to judge of the morality of a country.

I am absolutely ignorant, and shall never know, whether the crimes on which the tribunals have to pass judgment form the sixth, or seventh, or any other part you will, of the total number of crimes. What is important for me to know is that this ratio does not vary from year to year. On this hypothesis I can judge *relatively* whether one year has produced more or less crimes than another.

I can even compare the provinces together, admitting always that the ratio remains invariable not only from one year to the next, but also from one province to another, and that justice throughout pursues criminals with the same activity. In one particular kingdom, especially in provinces which have much similarity, these comparisons may be established without much inconvenience; but perhaps it is not so with provinces very distant from one another, and of very different manners. There would be few chances of the invariability of the ratio.

Comparisons become much more difficult when they relate to countries absolutely different. The elements then no longer present anything comparable, and we find ourselves reduced to hypotheses more or less hazarded. This has been too often lost sight of by authors, who have compared different people in respect to their morality, and who have only made use of their statistical documents translated into numbers.

The preceding may also show why, when considering morality, we must rather regard the number of crimes than the number of condemned. When a crime is proved, although an acquittal takes place, it is not the less established that there was a crime.

In recapitulation, we must admit that statistical documents only express more or less approximately the real values of the information which they should supply. Sometimes these documents represent with a high degree of approximation the figures they replace; such are those of births, deaths, and marriages in Belgium. Sometimes these data are only known with more or less extended limits, as the births and deaths in Russia, Turkey, and Greece: we must therefore be circumspect in using them.

It may also happen, as in the instance of criminal statistics, that the known values may be much inferior to the real values. It is impossible in this case, at least if there exists no determined and constant ratio, to supply the second by using the first.

LETTER XLIII.

INCOMPARABLE elements have very often been compared with one another in statistical works. By such means most absurd results may be obtained. I shall endeavour to show this by some special examples.

I find myself actually in Paris, and congratulate myself on the circumstance, for one of my friends, in a patriotic transport, has demonstrated to me mathematically (as it is the custom to say) that the capital in respect of morality is making such progress that soon there will be no more crimes committed in it. It is true that another friend, who also collects statistical facts in his leisure hours, has demonstrated to me, with a formidable profusion of figures, that certain streets in Paris are a prey to a mortality which far exceeds that produced by the most disastrous plagues; but by way of consolation he shows me that the Boulevard, on which I have taken up my abode, presents such a salubrity that he could almost guarantee immortality to those who dwell there. And by "immortality" I do not here mean that chimera of which poets and artists dream, but that most happy privilege of escaping the torrent of time which devours all, and of sitting upon its banks the peaceable spectator of the miseries and wrecks of others.

The demonstrations of these fair results well deserve to be reported. They may furnish the secret of other propositions, no less extraordinary, which have been proved by the aid of statistics.

Let us commence with the immortality: this is so precious an advantage that I think it ought to have my first attention. My neighbourhood numbers 300 inhabitants: two have died in the space of a year. The mortality then is 1 in 150 (I speak after my friend the statist) : if there had been but one death, the mortality would have been 1 in 300! This of itself is a fair result; but if there were no death at all (and such a thing is possible), he must needs then conclude that the part of the Boulevard where I lodge is the dwelling-place of immortality.

I should have good reason to congratulate myself on my abode, if there were not in its immediate neighbourhood a street which furnished matter for much reflection. There the mortality is 1 in 10,—that is more frightful than in many hospitals (it is my friend again who has proved this by his figures); for there have been during one year two deaths as on my Boulevard, but the street only counted twenty inhabitants! What is to be said? what thought? the figures are unhappily very exact, and the calculation no less so. Is there then so great a disproportion between my Boulevard and the unfortunate street of which I have been just speaking? Let us before answering wait until the end of next year, and perhaps matters will be entirely changed. Thus it is that we may err in operating on too small numbers, and in comparing them when they are not freed from the effects of accidental causes.

The mistake which I have just pointed out may appear gross; nevertheless, if it be desired to guard against it, a great number of books present striking examples of it. I would especially quote works on medicine, could the quotations be useful. We scarcely ever inquire what confidence a result merits: and the principle most often lost sight of is, that, all things being equal, the precision increases with the square root of the number of observations; and when the observations are not sufficiently numerous their use should be avoided.

This is the place to mention other sources of error which those who have written on population have not always avoided.

It must not be forgotten that the absolute mortality is calculated from two elements, which are far from being ascertained with all the precision desirable. In Belgium, for example, the number of the population is very badly determined : no census has been made for nearly thirteen years. I have reason to think that the number of the population is about one-tenth below its real value, and I have endeavoured elsewhere to prove this.* Were a well-made census to confirm my conjectures, it would result that in 1842 the number of the population would have exceeded 4,550,000; and the mortality would only be 1 in 44·3, or even 1 in 46·8, if the stillborn be disregarded, whilst it was valued at 1 in 40·5.

Another cause may have contributed to the belief that the mortality in Belgium is greater than it really is. The local administrations generally include among the deaths those of persons who, although foreign to the commune, may have died within it; as well as those of persons belonging to the commune who have died elsewhere, and whose registers have been transcribed in virtue of the 80th act of the civil code. These double entries must have exaggerated the number of deaths. Measures have been taken that, from and after 1842, such errors shall not recur. All deaths will be counted in the communes in which they occur: transcriptions will be entered separately.

Thus two principal causes have concurred to make the mortality in Belgium appear greater than it really is. The number of deaths is probably too high, and that of the population too low. Great prudence then must be exercised when we wish to compare the mortality of this kingdom with that of other countries, where the chances of error may be as great and incline in a contrary direction. France is undoubtedly the kingdom with respect to which these comparisons may be made with the least inconvenience: the deaths there are verified with the same rigour and under the influence of the same penalties as in Belgium. There is, moreover, reason to believe that the valuation of the population is there also too small.

It is not thus with England. The registration of deaths there is in some measure optional, and exhibits a number which should be much below the real value; so that the mortality there is represented by too small a number, supposing the element of the population to be exact. What must then be said of the comparison which it is wished to establish, with respect to mortality, with other countries whose population is ill known, and in which deaths are not inscribed according to any regular mode?

* Mémoire sur les contingents des milices, tome i. du Bulletin de la Commission Centrale de Statistique de Belgique.

Statists should carefully make known the means which they have employed in collecting numerical documents, and the precautions they have taken to render them comparable. In the absence of such information, we every instant see calculators obtain most divergent results in respect of the same things. It is time to put an end to such a state of things.

These remarks have nearly made me lose sight of the second demonstration. I have to explain how, after having had a foretaste of immortality, I have been able to conceive that France will one day be without crimes and without criminals, and that the same will be the case with Belgium. This is the demonstration in two words. Belgium before 1830 only acquitted 16 persons, out of 100 presented to it under accusations of all kinds. Since that period the number of accused has not augmented; but of 100 accused, instead of 16, the jury acquits 32 or even more. If this continue a little longer, perhaps none will be condemned,—an evident proof, my statistical friend tells me, that criminals will become unknown in Belgium.

I will allow myself to quote one last example, to show with what circumspection we should proceed to render results comparable one with another, which is the principal aim of statistics. Isolated figures have in fact no value of themselves: they acquire a value only as they are compared with other figures, in order to draw conclusions from them; but in these kinds of mathematical syllogisms it is of the greatest importance to prove well the premises.

The part of statistics in which, without any doubt, the most errors have been accumulated is that whose object is to prove the morality of a people. So I do not fear again returning to the subject, in the hope of throwing some new light upon it.

The word "crime" is very vague. What is reputed crime here is perhaps not so reputed on the other side of the frontier. This is a primary difficulty, when a comparison of one country with another in this respect is desired.

By confining ourselves to the limits of Belgium, we may eliminate this first cause of error. But supposing crimes to be so well defined as to leave no doubt, we never know, I repeat, but a portion of those which are committed; and, again, we only include in criminal statistics those which have been brought to justice. For the purpose of comparison, it is then not only necessary that the two countries should be under the influence of the same legislature, but also that all crimes should be proved in the same proportion.

This last difficulty is very great: however, it may be removed, as I have shown in my last latter, if we may be allowed to consider that the activity of justice, in proving crimes and prosecuting their authors, is the same in the two countries compared. I have endeavoured to show, in fact, that constant ratios are established between these three things,—the general number of crimes committed, the number of crimes known, and the number of crimes prosecuted. So that, without knowing the total number of crimes committed, we may nevertheless judge of the *relative states* of criminality.

There must then be the same legislation, the same repression, and the same activity of justice to discover the guilty. But this last element may vary from province to province: it should vary still more if, as is generally done, we compare the number condemned instead of the number accused, since we cause the introduction into the comparison of a new element essentially variable according to times and places. The intervention of the jury in criminal affairs necessarily renders absurd the comparisons established between the condemnations which took place before and those which have occurred since its introduction. We know, in fact, that the establishment of the jury in Belgium has doubled the number of acquittals.

LETTER XLIV.

ON THE USE OF STATISTICS IN THE MEDICAL SCIENCES.

NOTHING has been more strongly contested than the utility of statistics in the medical sciences; and from the manner in which they are applied, it should be so.

When one of those plagues which carry destruction among men, and which seem destined to cause anxiety among medical societies, where they leave the last traces of their ravages, bursts out, some doctors follow docilely the lessons of their predecessors, and others try adventurously new ways, either with a view of general utility or ,for a private purpose, and in order to attract public attention. All indistinctly collect statistics; but some confide their results to their memories, others to paper ; some even collect their statistics unwittingly, like M. Jourdain does prose. Those who have obtained the least success are cautious of speaking of their misadventures. There remain then but those who have succeeded or who think they have succeeded better than their fellows.

Among them, there are a very great number who only owe their successes to accidental causes, and who very probably would have less wherewith to flatter themselves if they had been called upon to exercise their art on a larger scale. But there are also some who only owe their success to their science and sagacity. Now it is here that doubt generally arises.

A medical man attributes the malady to a particular cause, and he has reason to do so in so far as concerns the maladies which he has treated. A second medical man attributes the malady to another cause, and he also has reason in like manner. But both are wrong, in that they each recognised but one single cause, while there really existed many. They should not have generalized what was but the result of particular cases: each one has only known one and the same face of the die which has shown itself many times in succession, and has not been able to discover all the other faces for want of a sufficient number of trials. Their contradictions only proceed from having incomplete notions.

It is these contradictions which strike the vulgar, and thenceforth he proscribes the figures from which have been deduced results which contradict or seem to contradict themselves. Let us add, for truth's sake, that frequently the desire to prove a success makes the memory less faithful in registering unfortunate cases, or they are thrown out, under one pretext or another, without any intention of insincerity.

But let us leave the judgments of the vulgar, and return to our example. The doctors will excuse me if I retort upon them a little of the scepticism which they generally use pretty freely against statists. If I suppose that they treat at hazard, and without doing good or ill, it will follow from the law of possibility, that the greater number will lose but few patients. Some will lose many and keep silence: others will save many; and it is generally these who will raise their voices. In stating the results of their practice, and I think them sincere, they only establish a fact which I willingly accept. Where then is the evil? It is found in their suddenly abandoning statistics, to throw themselves into the domain of conjecture.

They do not confine themselves to stating, "I have saved many that were ill;" but they add, "It is because I discovered the cause of the malady that I have learnt how to apply the true remedy." But they do not in any way prove the connection which exists between the effect and the pretended cause : this, however, is what they should do.

They even go further ; and, after having assigned a pretended cause to a disease, they treat conformably new patients who present themselves, without in most instances considering their constitution, their age, or their sex. It is here we find the abuse of statistics, if indeed it be statistics.

What renders the progress of medicine so slow and so uncertain is that the phenomena observed almost always depend on an infinity of causes, and that consequently they are scarcely ever perfectly comparable one with another. Nothing better shows the difficulties of this science than the delays it has experienced in its progress, notwithstanding the persevering labours and the superior genius of the great number of men who have been engaged upon it from remotest antiquity.

If a little attention be accorded to me, I shall be able to show the real difficulties which fetter the progress of medicine, and what are the aids which statistics may furnish towards surmounting them.

A doctor treats a patient, and cures him. He is afterwards called to a second patient, who is in identically the same circumstances as the former, who has the same constitution, the same age,—who is, in fine, exactly comparable in all respects. He will naturally apply the same therapeutic means which have already succeeded with him; and he may regard the cure as certain, if it be true that the same causes produce the same effects.

If there were a rigorous identity in all men, then one disease well observed, and followed by a cure, would be sufficient to obtain the same success every time that the same disease was reproduced in other individuals. But this perfect identity will perhaps never exist: we should suppose so at least, when we consider how much individuals may differ in respect of age, sex, and constitution by previous maladies, and an infinity of other causes. A doctor, during the whole course of his life, will not perhaps act twice under circumstances absolutely the same.

Thus, not to wander too far from my first hypothesis, I suppose that there is no difference in the human race beyond that of sex. Already the doctor will be less sure of his fact; and if he has succeeded in curing a man at first, perhaps he may fail in the treatment of a woman. He must have recourse to experiment, and prove whether the means employed in the first instance will again succeed. He requires then at least two observations.

If there were not only difference of sex, but a difference of age in addition, the number of observations should be further increased. Thus, supposing (in order to simplify the matter) that there are only three classes—infants, adults, and old people, each composed of individuals of exactly the same age, there would exist but six kinds of individuals; and there should be at least one observation for each of the six different cases which might present themselves. However, as ages may vary by all the several gradations, should there not be as many observations as there are possible ages? The number would be infinite.

What then must be the case if we have not only to take account of sex and age, but also of all individual peculiarities? I repeat that the whole career of a doctor would not suffice to furnish an opportunity of observing two patients in exactly the same circumstances.

Such, it appears to me, is the strongest objection that can be made to the use of statistics in the medical sciences. If it were wished to foresee all the cases which could be presented, and to collect sufficient observations to verify all possible combinations, we must despair of ever arriving at anything satisfactory,—we must not only renounce the use of statistics, but also of observation. Experience, in fact, would be but a vain word, since one single disease might be subject to an infinite number of modifications, under the influence of all the causes to which it owed its rise.

Let us see, however, why doctors, even those who have most eagerly repudiated statistics, have not entirely despaired of the future of their science. It is because they have felt that there does not exist a great number of essentially different cases, although the causes which may influence one disease are in general very numerous and susceptible of variation in an infinity of degrees. In many diseases, for example, the difference in sex produces no appreciable effect,—it is the same with a slight difference in age: so that these causes may be considered as exercising nearly the same degree of influence. It only remains then to discover the causes which show a very decided individual influence, and to ascertain from observation the degree of energy which belongs to each. This examination can only be performed by men of an exquisite tact and of a sure judgment,—by observers endowed with that aptitude for patience which Buffon called genius. Now, could we believe that men of this degree of prudence would keep no account of their observations, or that, if they kept account for the purpose of subsequent comparison, that they would prefer to trust to their memories rather than commit their observations to paper. But so soon as these observations are collected, so as to render them comparable and to draw inferences from them, statistics have been formed. There is but this difference between those who write down the results of their observations and those who entrust them to their memories, that the former conform to the principles of science, while the latter palpably fail to do so.

In the medical sciences, moreover, and particularly in regard to public health, all facts are not equally complicated, and consequently do not present the same difficulties to a statistical analysis. When the observer has discovered them, he is certain of being able successfully to apply the method founded on the calculations of probabilities.

When Jenner published to the world his important discovery, it was at first felt that, in order to ascertain the value of vaccination, the facts should be registered and compared with the most scrupulous attention, having regard to all the causes which might oppose its efficacy. This purely statistical method placed beyond doubt the benefits of vaccination. It was thus, again, that it was, at a later period, proved that this precious preservative had in certain circumstances but a temporary action.

Let us take another example, which will show us how we may for a long time err, from our repugnance to counting, in questions which are exclusively within the domain of numbers. The science of statistics even before it had a name interfered in medical questions. The progress of the pulse at different ages had been determined, and the part it had to play in either sex recognised. Only the observations, whether bad or too few, had induced an error respecting the pulsations in the old. The mistake was successively repeated in all treatises on Physiology; and although doctors were every day feeling the pulses of their patients, and acting on their indications, the thought never occurred to them to verify the fact. It was not until these latter days that it was proved, by new observations, that the pulse, contrary to the received opinion, is generally quicker in the old man than in man in his prime. Instead of disdainfully repelling the aid of statistics, would it not be better to make use of it, in order to destroy errors existing in works consecrated to the medical sciences? The greater part of the numbers met with there require a severe revision; and we should perhaps be astonished at the errors which would require correction.

Statistical data would, doubtless, be of the greatest utility, if they were always collected dispassionately, and without the intention of making a preconceived idea prevail. When a surgeon prefers one kind of operation to another, it is because he finds in it more chances of success. But to arrive at this conclusion he must have counted and compared, he must have had recourse to statistics. The deplorable abuses which have been made of this science have often led to a belief that its use is impossible. When in an individual the existence of stone has been proved, there are different modes of operating to extract it. Which must be preferred? It is evident that this question can only be decided by collecting observations made conscientiously and with discern-ment. If, all things being equal, lithotrity saves more patients than incision, it must be employed. I have said intentionally *all things equal,*—that is to say, the operators having the same dexterity, all the cases observed being exactly comparable. Here is the difficulty. I even lay aside the appreciation of the pain resulting from either method. We must know the influence of sex, of age, of the constitution of the patient, of the period of the malady, &c. To take account of all these influences would require a considerable number of observations made with precision, and discussed with intelligence. But this discussion can only take place by statistical methods, which are those of every science of observation: to desire to dispense with them is to throw ourselves into a blind empiricism, and to reject the elucidations of experiment.

A wise discussion of facts teaches which of the two operations compared has presented the greatest success, whether this success has been obtained in a general way, or more especially in respect to one sex or a certain age, or any other particular circumstance. All reasonable men will I think agree on this point, that we must inform ourselves by observation, collect well-recorded facts, render them rigorously comparable before seeking to discuss them with a view of deducing their relations, and methodically proceeding to the appreciation of causes. In lieu of this, what do we see? observations, incomplete, incomparable, suspected, heaped up pellmell, presented without discernment, or arranged so as to lead to the belief of the fact which it is wished to establish; and nearly always it is neglected to inquire whether the number of observations is sufficient to inspire confidence.

I insist especially on this last point, because in the art of curing there are questions influenced by so many different causes that it will perhaps never be possible to obtain a satisfactory solution of them. I go further: even supposing that there were exact solutions in the most complicated cases, I say they will be of no use in application, although they may be very useful for the public hygeine. They will only have a general value; and it would be absurd to apply them to individuals, because account cannot be taken of all their peculiarities. We might as well seek, from a table of mortality, at what age a particular individual should die. No one, however, doubts the utility of tables of mortality in medical researches and in speculations on the duration of life. Statistics, or rather methods of observation founded on calculation, will have already rendered a true service if they have examined the most simple cases, those in which few influencing causes act.

Has the science of statistics then rendered such poor services, in teaching us the influence exercised on deaths by ages, by sexes, by profession, by atmospheric circumstances? in studying that which relates to births, to the number of accouchments and of stillborn, to vaccination? &c. Among the facts it has proved there are many of great importance, which have been submitted to the examination of masters of the art, and which remain unexplained. I will quote in particular the greater mortality of male children at the time of birth. Whence does it proceed? Has it been sought to account for the peculiarities relating to the stillborn, and to combat the causes which in certain circumstances swell their number in so deplorable a manner?

Statistical questions relating to surgery are in general more easily attacked than those which relate to medicine: in the one the malady is seen, and in the other it must almost always be divined. The doctor must establish a sort of inquest, and take, in the interrogatory to which he subjects the patient, the same precautions that the judge employs with regard to the criminal from whom he would obtain his secret.

There are two kinds of difficulties. By the side of the question of the therapeutics is placed that of the diagnostics. The cure of the patient depends on a compound event,—that is to say, 1st, that the doctor will discover the disease; 2nd, that he will know the mode of treatment to be adopted. Now the second question generally receives attention before the first; and I think wrongly. I will on this subject quote some very just reflections: I extract them from a small work which I have just received from a friend, who knows how to embrace with a philosophical glance many sciences of observation, which he cultivates with an equal success. " Therapeutic questions," he says, " are not those with which to commence. I would have calculation first applied to questions of diagnostics, or the distinguishing of symptoms. Such symptoms being given, what is the probability that the patient is affected by one malady rather than another? the probability of its duration, of its return, and of its passing into a chronic state when it is abandoned to the power of nature only? These different questions and many others once resolved, I would then pass to the study of the influence of medicine."

In this last study, to judge with knowledge of the cause of the advantages which therapeutics may present, we must commence by inquiring what would become of a man affected with such a malady if abandoned to the force of nature only. Perhaps we might be led to conclude that, in doubtful and difficult cases, it is better to give up the patient to the efforts of nature than to the remedies of art, confining ourselves to the use of a careful diet. Different kinds of treatment have less influence on mortality than is generally supposed. A respected and learned man, Doctor Hawkins, thus expresses himself: " A friend took private notes on the comparative mortality under three doctors in a hospital. The one was *eclectic*, the second pursued the *expectant* system, and the third the *tonic* regimen. The mortality was the same; but the duration of indisposition, the character of the convalescence, and the chances of relapse were very different." Thus the mortality was the same. We might draw the same conclusions from the documents collected in the principal hospitals in Europe: the mortality varies between very narrow limits, and depends more on the principals of the hospitals than the therapeutic means employed. From whence it would appear to result that administrative science had at least as much influence as medical science; and we conceive it should be so. Of what use is it to call in the best informed doctors, if their prescriptions are not followed, and if, during their absence, ill-understood attention or imprudence destroys all the good that might have been effected?

Did I not fear being taxed with exaggeration, I would say that a good administration perhaps saves more patients in hospitals than the science of the most skilful doctors. To judge of its influence on the health of men in vast establishments, let us examine what is passing in prisons: the mortality will there be seen to vary between very wide limits. Without even going out of Belgium, I find that in the house of correction at Ghent the deaths were proportionally less numerous than in the privileged classes of society; while in the prison at Vilvorde there reigned, during the years 1802, 1803, and 1804, such a mortality that never were men during the most frightful plagues, never were soldiers during the most destructive wars, decimated in a more frightful manner. Annually, three prisoners out of every four died. This scourge, produced by a vicious administration, began to punish with less intensity in 1805 (thanks to useful reforms); and two years afterwards everything had returned to a normal state.

It would be impossible to present your Highness an example more instructive, and which more merited to fix your attention. A science whose mission it is to reveal such facts cannot be without importance in the eyes of an enlightened prince.

LETTER XLV.

ON THE USE OF STATISTICS TO THE ADMINISTRATION.

STATISTICAL documents have a double interest : they are useful alike to the sciences and to the administration. It is only by consulting the past that the statesman can form just ideas of the future, discover whether a country possesses the elements necessary to realize with success projected plans, appreciate which are the laws which require reform, and throw light on a crowd of important questions.

The great extension of the railway system which has taken place in Europe gives rise to many political problems, of which we are far from possessing the elements of solution. It will be difficult to foresee henceforward the variations which will take place in the populations of towns, in the price of lands, in the principal seats of the different manufactures, and in all its social transactions generally. It will be difficult to believe one day that a simple acceleration in the transport of travellers and merchandise could have had such consequences. Human life is lengthened : the globe itself seems to have changed its dimensions ; towns have drawn near to one another: already our own Belgium seems in some measure enclosed within the precincts of its capital. Civilization extends its level from one extremity of Europe to the other: the characteristic and the picturesque is being effaced in every people ; whilst locomotives daily make new breaches in the barriers of the Customs, until it shall entirely have destroyed them.

Statists should be eager to register, from this time forward, all the facts which may assist in the study of this vast transformation in the social body which is in process of accomplishment. No one can completely foresee the consequences, and yet each one seeks to appreciate them.

But, without considering things under so general a point of view, how many statistical questions of actual interest there are connected with the establishment of railways ! Questions relating to receipts and expenditures, to the duration of materials, to the scale of fares, to the tariff for merchandise, are so multiplied and so important that it would be to the interest of an enlightened government, in order to obtain solutions to them, to submit them to some educated men who would make it their exclusive study.

Only to quote one example. There is a dependence between the number of travellers transported each day and the fares they have to pay : this dependence is such that the receipts augment or diminish according to the scale of fares. Every one can conceive, in fact, that if the fares were too low, the number of travellers, although more considerable, would not be sufficient to pay the expenses of the enterprise; if, on the contrary, they were too high, the number of travellers would diminish, and the administration again would run the risk of a loss. There is then a *maximum* which can be obtained, and which can only be determined by the aid of good statistical documents.

Nevertheless, the question is more complicated than it may appear; for the government not only receives a sum paid in the railway offices,—it levies also a kind of indirect contribution on the expenses, and on all the transactions occasioned by the journeys. The general activity—the increase of vitality which the whole country receives—gives a new energy to commerce and industry, which are, with agriculture, the chief supports of a state.

A trial connected with this kind of question has been recently made in England. The government has suddenly reduced very considerably the postage of letters. What has been the result? The deficit in the receipts of the Post Office were at first very considerable; but the number of letters having afterwards progressively augmented, it is found definitively that the revenues have nearly reached their ordinary amount: the difference which yet exists is, doubtless, more than compensated by the indirect advantages which the government and private individuals have derived from it. The results of this administrative measure, so skilfully combined, could have been immediately verified by statistical documents collected with care.

It is not sufficient, however, to have appreciated the material advantage of an innovation. The moral advantages are sometimes of a greater importance : they are too often lost sight of. Belgium has found in the establishment of its railways a moral power much superior to the pecuniary benefits which may result from its vast enterprise. The government, emerging from a revolutionary crisis, was able to give a new direction to people's minds —to lead them back to habits of order and work; and foreign nations, who had ill judged us at first, and who thought us given up to anarchy, had faith in our future: they could not refuse confidence and esteem to a people who, on the morrow of a revolution, courageously undertook gigantic works before which other states founded on the most solid bases fell back.

A government in modifying its laws, especially its financial laws, should collect with care documents necessary to prove, at a future date, whether the results obtained have answered their expectations. Laws are made and repealed with such a precipitation that it is most frequently impossible to study their influence : the laws of import and export duties present melancholy examples. It is with them as with the postage in England. There are prices which we must learn to attain: they are those which best conciliate all interests.

When taxes are too high recourse is had to fraud, or the imports diminish. There exist, on the contrary, numerous examples which prove that by lowering the Customs or Excise charges the revenues of the Treasury have been considerably augmented.

Most of the civilized nations have very inexact ideas of matters which it is most important for them to know. I will mention in particular the value of harvests, and of the principal articles of consumption. However, it is indispensable to know when laws are established whether a country can rely upon itself, and to be aware, on the eve of a crisis, of the extent to which it must be provided for from abroad.

A wise administration should also enter among its statistics information which it may suddenly want in extraordinary circumstances, such as a declaration of war. It is important to know whence the provisions necessary for the men and horses may be procured, where lodgings may be found; whence to take the carriages and horses necessary for the transports.

I think I need not further insist upon the importance of a good census and of a well-kept record of the civil state : I have said enough in my preceding letters. All questions which are connected with population deserve, in general, the greatest attention on the part of the government. There is one which from day to day offers more interest, and which seems to be still left in the shade, although it has already excited much uneasiness in some states. For that reason I will submit it: it refers to the legitimacy of births. It is a well-known principle that under the influence of the same causes the effects remain the same. If, for example, illegitimate births increase progressively, the cause must be sought for somewhere. In many great cities, such as Paris and Brussels, one-third of the births are illegitimate. At Munich the ratio is greater still, and there are counted as many illegitimate as legitimate. What can be the cause? Must it be attributed to a legislative measure designed to obviate other evils, and be said that the prohibition to contract marriage without being able to guarantee the present and future support of the wife causes concubinage. A perfect knowledge of the matter only can resolve this important question.

The statistics of crimes, as well as the statistics of births, has already introduced useful reforms in the laws. It has shown, for example, that when the disproportion between the crime and the punishment is too great, the latter is not applied, and consequently remains almost useless. It is thus that the punishment of death attached to infanticide is rarely applied: to give the law its full effect, its rigour must be modified.

When a penal law is modified, the influence of the change should be felt : if this influence leave no trace, and the results of previous years continue to recur, the modification is without effect, and consequently illusory. When, on the contrary, the effects produced are marked, they teach whether the modification has been advantageous or injurious.

France gave the first example of the statistics of crime carefully collected on an extended scale. From the first publication, there might be read a result so striking that I did not hesitate to proclaim it, although many were incredulous of it: it is, *that there is a budget which is paid with a frightful regularity,—that of the prisons, the galleys, and the scaffold;* and I added, *It is the former that we must endeavour to reduce.*

Some persons, at first, only saw in the expression of the fact revealed by the statistics of crime a tendency to materialism; and, in their prejudice, they did not even dream of the sense I might attach to the last words.

But it is precisely in the regular return of the same effects under the influence of the same causes that the legislator finds the most consoling idea, and the proof that by advantageously changing a law he will necessarily produce a useful effect on the future of the nation. For what end should he make his reforms, if he were not certain that they would produce fruits, and that these fruits would be durable? He could not flatter himself to cause the disappearance of all the crimes which pollute society; but it may be conceived that there is a collection of laws, an enlightened administration, and a social condition such that the number of crimes may be as much reduced as possible. This last amount depends on the internal organization of man, and the excess is in some measure the produce of social organization.

I may be allowed to quote another example of the utility of criminal statistics. I have already made it known elsewhere; and I do not fear to repeat it, because it seems to me as curious as it is conclusive, and because in this circumstance I predicted the future without my predictions being falsified by experience,—a thing rare enough for me to be allowed to be somewhat proud of it.

When, in 1826 and 1827, the first statistics of the tribunals of France and Belgium appeared, my attention was turned to the repression,—that is to say, the ratio of the number of the condemned to that of the accused; and I printed as follows in the Belgian statistics the first results of which I had to publish.*

* *Recherches Statistiques sur le Royaume des Pays-Bas,* 1 vol. 8vo., 1829, Tarlier; and *Mémoires de l'Académie de Bruxelles,* vol. v., 1828.

" In 1826 our tribunals condemned 84 individuals out of 100 accused, and the French tribunals 65 : the English tribunals have also condemned 65 per cent. during the last twenty years. Thus, *out of* 100 *accused,* 16 *only have been acquitted with us, and* 35 *in France as in England.* These two latter countries, so different in manners and in laws, however pronounce in the same manner on the fate of the unfortunate submitted to their judgments; whilst our kingdom, so similar to France by its institutions, acquits a half less of the accused. Should the cause of this difference be sought in the fact that we have not the institution of the jury, which our neighbours have ? We think it is so.

" Let us examine, in fact, what is passing before the correctional tribunals where the judges only give sentence, as in our tribunals. We shall find in France the same severity as with us. Of 100 accused, only 16 are acquitted. Let us examine the tribunals of police simply,—the same severity : of 100 accused, only 14 are acquitted. The preceding will lead us then to the conclusion that *when* 100 *accused come before the tribunals, whether criminal, or correctional, or simple police,* 16 *will be acquitted if they have to be dealt with by judges, and* 35 *if they have to be dealt with by a jury.*"

Such were the conclusions I came to from the first statistical documents on crime which were published in France and Belgium. I did not then know that the following year would realize my conjectures in the most brilliant manner. The revolution of 1830 detached Belgium from the Kingdom of the Netherlands and gave it the institution of the jury. Immediately the acquittals took the same course as in France.

The chances of acquittal for one accused were then doubled in Belgium by the sole fact of the institution of the jury ; and of 100 accused, 16 who would have been condemned by the system in operation anterior to 1830 were returned to society. Is this a benefit ? Is this an evil ? I confine myself to giving over this remarkable fact to the meditation of the legislator.

It is true, it might be asked if the accusations were always made with the same regularity; for on this preliminary judgment would depend in a great measure the value of the repression.

Statistics, I repeat, offers one of the surest methods of appreciating the efficacy of the laws. At a certain time the condemnations for forging bank-notes were very numerous in England, and nevertheless they incurred capital punishment. Instead of continuing to punish the guilty, they applied themselves to the introduction of reforms in the manufacture of these notes ; and immediately afterwards the number of condemnations was found considerably reduced. Such a measure taken some years earlier would perhaps have saved the life and the honour of many an unfortunate.

Are any other proofs necessary to show the wise circumspection which should govern the legislator, and the useful lessons he may derive from statistical documents ?

LETTER XLVI.

ON THE ULTERIOR PROGRESS OF STATISTICS.

STATISTICS, even to the present time, has been received with great favour. Many learned men have shown the resources that might be drawn from it, and have presented solutions to questions equally new and instructive. But as the imagination and unenlightened zeal of different writers have exceeded the limits which should have been imposed on them by observations as yet too few, and moreover too inexact, which we possess, distrust has succeeded to the charm they at first presented.

If anything however should astonish, it is that, with so few resources as the science possessed, it has been enabled already to establish so great a number of facts interesting to society. It is true frequent mistakes have been committed; but they have been successively discovered and signalized,—and perhaps even these errors have had their use.

In some countries figures have been accumulated by an immoderate desire to contribute to the progress of statistics, and to throw light on the different movements of the state. This profusion of tables, for the most part inexact, has only encumbered the domain of the science with materials inconvenient and often injurious.

It is important above all things, as I have not ceased to repeat, that the statistical documents which are published should be exact, comparable, and present all the necessary guarantees. But, in the actual state of things, comparison cannot be established even in the limits of a single kingdom. Every administration publishes its documents without putting itself in harmony with its neighbours. We often find different figures employed to express the same things, and nearly always dissimilar classifications when the most rigorous uniformity is necessary: this is especially remarked in classifications by ages, in dividing the population into different professions, in the nomenclature of diseases, and in that of crimes made known to the tribunals. In France, while the minister of commerce values corn by hectolitres, the administration of the Excise counts by metrical quintals and kilogrammes.

These disparities are so many obstacles to the progress of statistics: they have struck all those who have been engaged on this science practically. The celebrated Malthus expressed to me one day his regret that there was not any country in Europe in which statistics were organized in a manner to respond to the requirements of the science. The sacrifices that might be made to obtain so desirable an end would certainly be well compensated by the works which it would elicit from the most enlightened men: means of amelioration, if there were such a country, would become the object of their constant meditations. Belgium appeared to the eyes of this celebrated English economist to combine the most favourable conditions for this purpose. Placed between three principal nations of Europe—France, England, and Germany,—constantly traversed by the travellers of the different countries,—it offers to each easy means of taking personal cognizance of the different places, and of checking the statistical documents. Its small extent renders these verifications easy. They have besides guarantees of exactness in its well-managed civil state, in an administration established on good principles. There are to be found there—agricultural districts and purely manufacturing districts, mountainous lands and plains. The composition of the kingdom allows of all kinds of studies.

Malthus at my request penned a note, which I took care to remit myself to the Belgian Government; but this writing was probably lost sight of, and underwent the fate of the many projects which sleep in the ministerial portfolios, waiting for the day when they shall be again brought to light. However, an enlightened statesman, M. Liedts, has since realized (in part at least) the ideas of Malthus, in creating here the Central Commission of Statistics, which has ramified and covers the whole of Belgium with a vast net adapted to favour any inquiries the government may think fit to make. It is to be desired that this unity should extend itself further.

When two different countries are concerned, it seems to have afforded pleasure to make every kind of comparison impossible. It will yet require much time to put such a chaos in order.

Some governments have, however, felt the inconvenience of such a state of things, and have sought to put an end to it. Belgium and Sardinia have first entered upon this path: other civilized states will in the end follow their example. To France and England are due special statistics of great interest. Is it not to be regretted that these two homes of civilization have not yet adopted measures necessary to proceed with unity in the editing all the documents they publish, and to avoid the double entries which are so often met with in them?

When an uniform course shall have been adopted by each state, there will remain one last step to be taken. This will be to make the different publications uniform, at least as far as possible, by reconciling the general interests of the science with the particular interests of each country.

The force of circumstances is every day tending to the establishment of this desired uniformity. For when classification of the statistical documents of a country has been judged to be good, it is naturally sought elsewhere to differ from it as little as possible. But the desired end will be attained in a more sure and rapid manner by the establishment of grand centres of action, which will put themselves in direct relation with one another.

It is in such an organization that the future of statistics reposes. What is essentially wanting are good observations which are comparable together. I think I have sufficiently shown that scientific methods to put them in operation will not be wanting. Isolated man sees his actions enclosed within too narrow a circle for him to dream of collecting all the materials to compose the edifice: he must, to ensure success, have recourse to the generous intervention of governments.

I am happy to be able to plead here the cause of a science which is dear to me, and which will need support to realize all that we have a right to expect of it. I could not do it under better auspices; for the princes of the House of Your Highness have always been the protectors of science and the friends of enlightenment: and I have been in the happy position of being able to judge for myself that these hereditary qualities have not degenerated.

NOTES.

LETTER III., page 12.

" To estimate the probability of the return of an event which has reproduced itself periodically several times in succession. *Divide the number of
times the event has been observed increased by unity by the same number
increased by two.*"

That is to say, that if we represent the number of times that the event
has occurred by a similar number of white balls which we throw into an
urn, adding also one other white ball and one black ball, the probability
of the reproduction will be equal to that of drawing a white ball.

Each reproduction of the observed event is equivalent, consequently, to
putting a fresh white ball into an urn in which there were before beginning
the trials one white and one black ball.

Page 14.

" To calculate the probability that an event observed any number of times
in succession will re-occur, this rule should be followed. *The probability
is equal in value to a fraction which has for its numerator the number of
observations plus 1, and for the denominator the same number plus 1, and
plus also the number of times that the event is to re-occur.*"

Making use of the above comparison, we consider it the same thing as if
each reproduction of the observed event corresponded to putting a white
ball in an urn where 'there were already, before commencing the trials, a
white ball and as many black balls as it is supposed that the event observed
should re-occur times.

When, for instance, an event has been observed *m* times in succession,
the probability that this event will be reproduced *n* times more is

$$P = \frac{m + 1}{m + n + 1}.$$

The probability is the same, as we have already seen, as that of taking a
white ball from an urn which should contain as many white balls as the

event has been observed times plus 1, and as many black balls as the number of times the expected event should reproduce itself.

If, in the preceding formula, we regard the quantities P, m, and n as three variables, we shall have the equation of a surface of the second degree. This surface is of the nature of those called "plane surfaces;" for we may apply a right line throughout its extent parallel to the plane of the co-ordinates of m and n. Indeed, making P constant, there remains only the equation to a straight line, which may be written

$$m = \frac{Pn}{1 - P} - 1,$$

whence

$$\frac{m + 1}{n} = \text{a constant.}$$

In order that the probability of the event observed m times, being repro-duced n times more, should remain the same, we must then have this very simple relation.

LETTER IV., page 16.

"In general, the probability that there exists a cause which necessitates the reproduction of an event increases much more rapidly than the proba-bility of the next occurrence of the event."

We will endeavour to demonstrate this proposition mathematically.

When an event has been observed x times in succession, we have as the probability of the reproduction of this event

$$y = \frac{x + 1}{x + 2}.$$

This equation belongs to the hyperbola of which this is the form,—

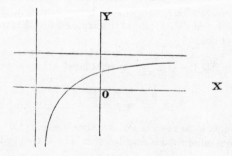

For $x = \infty$, $y = 1$,—that is to say, that after an infinite number of observations the probability becomes certainty. For $x = 0$ (that is, before beginning the trials) the probability is $\frac{1}{2}$, which characterizes the state of doubt on what is to happen.

As the probability that the event is not produced by chance, but is facilitated by causes, we have

$$y = \frac{2^{x+1} - 1}{2^{x+1}}.$$

This is the equation of the logarithmic curve. For $x = \infty$, we have $y = 1$: thus, after the indefinite re-occurrence of the event, we are sure that it depends on certain causes. For $x = 0$, we have $y = \frac{1}{2}$, which characterizes also the doubt in which we are before the trials as to the existence or non-existence of a cause favourable to the reproduction of the event.

The two probabilities converge then to certainty as the event is more frequently repeated, but with unequal rapidity. The latter probability increases the more rapidly.

In fact, for the same value of y, we have

$$\frac{x+1}{x+2} = \frac{2^{x'+1} - 1}{2^{x'+1}}, \text{ whence } 2^{x'+1} = x + 2;$$

or

$$x = 2^{x'+1} - 2 = 2(2^{x'} - 1).$$

This equation between the two variables x and x' is still that of a logarithmic curve.

THE PROBABILITIES ARE THEN EQUAL,

FOR THE EXISTENCE OF A CAUSE,	FOR THE RETURN OF THE EVENT,
After 1 occurrence.	After 2 occurrences.
2 „	6 „
3 „	14 „
4 „	30 „
5 „	62 „

In the work of M. Cournot, *Exposition de la Théorie des Chances et des Probabilités*, Paris, 1843, page 155, may be seen the observations which he has made against the formula

$$\frac{2^{x+1} - 1}{2^{x+1}},$$

which has been called " Bayes' rule."

Letter VII., page 26.

"It was at a gaming-table it was brought to light."

The following is from the discourse on the Life and Works of Pascal, by Bossut, author of the *Essai sur l'Histoire Général des Mathématiques*.

"The same principles gave rise to a new branch of analysis, which has since been very fruitful; and it is to Pascal again that the elements are owing. The Chevalier de Méré, a great gambler, but no geometer, had proposed on this subject two problems to Pascal. The one was to find in how many throws it might be expected to obtain *sonnez* (two sixes) with two dice: the other was to determine the lot of two players after a certain number of throws,—that is to say, to fix the proportion in which they should divide the stake, supposing they consented to separate without finishing the game. Pascal had soon resolved these two questions. He has not given the analysis of the former: we only see by one of his letters to Fermat, that from the result of his calculation, it would be disadvantageous to undertake to obtain in twenty-four throws *sonnez* with two dice; which is true in fact, as it is equally true that there is an advantage in trying the same thing in twenty-five throws. But he has left us, relatively to the second question, a writing to determine in general the allotments which should be made to two players who play several games; and he has again treated the same matter in his letters to Fermat. The Chevalier de Méré, who had resolved, with the aid of natural logic, some particular and easy cases of these problems, incapable of appreciating the researches of Pascal, but proud of having given occasion for them, thought right to run them down ; and carrying to excess the ridiculous liberty which most men of the world arrogate to themselves of judging everything, of improving everything, without having sounded the depths of anything, he dared to write to Pascal that *the demonstrations of Geometry are very generally false; that they prevent our entering upon higher knowledge which never deceives;* that they lose to the world the *advantage of remarking, in the countenance and air of the persons whom we see, many things which might be of much use,* &c. If this ridiculous letter has any sense, we see that the author regards the art of seizing the weaknesses of men, and profiting by them, as the supreme science;—the opinion of a vile and depraved mind, which no one would dare openly to announce. We feel that the judgment of the Chevalier de Méré on the discoveries of Pascal could only excite pity, and not indignation. Fermat, Roberval, and the other great geometers of the time, applauded these same discoveries; and their suffrage would have consoled the author had he had need."

LETTER VII., page 28.

"It is necessary that in every kind of game the mathematical chances of the players be equal."

Let p and p' be the probabilities of winning of the two players; s and s' the sums they respectively stake. We should have, in order that the game be equitable, the proportion

$$p : p' :: s : s'.$$

And we deduce from it that $p\, s' = p'\, s$. The quantity $p\, s'$ is the mathematical expectation of the first player, and $p'\, s$ the mathematical expectation of the second player. These quantities then are always equal, if the stakes were made in proportion to the probability the players had of winning.

In a lottery this equality does not exist. Calling s the sum exposed by the player, and s' the sum he expects to win, we find that s' is always less than the value required for the equality, and so much the less as the ratio of s' to s is greater. The following table will sufficiently show this: it is constructed on the basis which regulates the French Lottery.

NATURE OF THE GAME.	For 1 Franc staked, Sum which the Player		Loss on what he should receive.
	ought to receive.	does receive.	
Simple Drawing .	18	15	$\frac{1}{6}$
Determined do. .	90	70	$\frac{2}{9}$
Simple Ambe . .	400	270	$\frac{1}{3}$
Determined do. .	8,000	5,100	$\frac{3}{8}$
Terne	11,747	5,500	$\frac{1}{2}$
Quaterne	511,037	75,000	$\frac{6}{7}$
Quine	43,949,267	1,000,000	$\frac{43}{44}$

Letter XIV., page 60.

The numerical table contained in Letter XIV. only contains the coefficients of the binomial in its successive powers to the thirteenth. In general, if the probabilites of two simple events are p and p', the following development will give all the compound events possible m in number which can be formed of the simple events.

$$(p + p')^m = p^m + m\ p^{m-1} p' + \frac{m(m-1)}{1.\ 2} p^{m-2} p'^2 + \&c.$$

The repetition of the event m times in succession has for its probability p^m, and the coefficient indicates that this event is unique. The event composed of $m - 1$ times the first, and once the second, has for its probability $m\ p^{m-1} p'$; and so on. Putting $p = p'$, it becomes

$$2^m = 1 + m + \frac{m(m-1)}{1.\ 2} + \frac{m(m-1)(m-2)}{1.\ 2.\ 3} + \&c.$$

The different combinations in which the two simple events present themselves m times are 2^m in number.

Page 62.

"Using these several precautions, I caused 4,096 successive drawings to be made."

The balls were really drawn one by one, and, after inscription, were replaced in the urn. But the series of balls was afterwards divided into groups, proceeding one by one, two by two, three by three, &c.; which really presented the same results as if the balls had been taken by ones, by twos, or by threes, &c., supposing an infinite number in the urn.

The following table, in a double form, presents the results of these operations. I give first the results obtained by the successive drawings; then the results reduced proportionally to be comparable with the numbers calculated in the table in Letter XIV., page 60.

Table showing the absolute Number of White and Black Balls drawn from an Urn, taking one, two, three, &c. Balls at a Time.

Number of Balls in each Drawing.	NUMBER OF BLACK BALLS IN EACH DRAWING.													Number of Drawings.
	0	1	2	3	4	5	6	7	8	9	10	11	12	
1 by 1	2066	2030												4096
2 „ 2	543	980	525											2048
3 „ 3	185	518	474	188										1365
4 „ 4	69	268	370	246	71									1024
5 „ 5	30	125	277	224	136	27								819
6 „ 6	17	65	166	192	166	69	8							683
7 „ 7	9	34	104	151	148	95	40	4						585
8 „ 8	2	17	58	123	128	111	52	19	2					512
9 „ 9	2	7	28	86	89	113	80	34	5	1				455
10 „ 10	0	7	23	48	77	104	80	46	20	5	0			410
11 „ 11	0	3	13	35	58	78	84	55	32	12	2	0		372
12 „ 12	0	2	10	22	35	71	77	57	37	17	10	3	0	341

Table showing the relative number of White and Black Balls drawn from an Urn, by one, two, three, &c. Balls at a Time.

Number of Balls in each Drawing.	NUMBER OF BLACK BALLS IN EACH DRAWING.													Number of Balls Drawn.
	0	1	2	3	4	5	6	7	8	9	10	11	12	
1 by 1	1·008	0·992												2
2 „ 2	1·06	1·91	1·02											4
3 „ 3	1·08	3·03	2·77	1·10										8
4 „ 4	1·07	4·18	5·78	3·84	1·11									16
5 „ 5	1·2	4·9	10·8	8·7	5·3	1.1								32
6 „ 6	1·6	6·1	15·5	18·0	15·5	6·5	0·8							64
7 „ 7	2·	7·4	22·7	33·0	32·4	20·8	8·8	0·9						128
8 „ 8	1·	8·5	29·0	61·5	64·0	55·5	26·0	9·5	1·0					256
9 „ 9	2·2	7·9	42·8	96·8	100·1	187·1	90·0	38·3	5·6	1·1				512
10 „ 10	0	17·5	57·4	119·9	192·3	259·7	199·8	114·9	49·9	12·5	0			1024
11 „ 11	0	17	72	193	319	429	462	303	176	66	11	0		2048
12 „ 12	0	17	120	264	420	853	925	685	444	204	120	36	0	4096

LETTER XV., page 67.

			SCALE OF POSSIBILITY.	SCALE OF PRECISION.	SCALE OF POSSIBILITY.
SCALE OF POSSIBILITY AND OF PRECISION.					
GROUPS OF		Rank of Groups.	Probability of Drawing each Group. *Table A.*	Sum of the Probabilities, commencing at the most probable Group. *Table B.*	Relative Probability of Drawing each Group. *Table C.*
499 white balls and 500 black		1	·025225	·025225	1·000000
498 ,,	501 ,,	2	·025124	·050349	·996008
497 ,,	502 ,,	3	·024924	·075273	·988072
496 ,,	503 ,,	4	·024627	·099900	·976285
495 ,,	504 ,,	5	·024236	·124136	·960789
494 ,,	505 ,,	6	·023756	·147892	·941764
493 ,,	506 ,,	7	·023193	·171085	·919429
492 ,,	507 ,,	8	·022552	·193637	·894040
491 ,,	508 ,,	9	·021842	·215479	·865882
490 ,,	509 ,,	10	·021069	·236548	·835261
489 ,,	510 ,,	11	·020243	·256791	·802506
488 ,,	511 ,,	12	·019372	·276163	·767956
487 ,,	512 ,,	13	·018464	·294627	·731958
486 ,,	513 ,,	14	·017528	·312155	·694860
485 ,,	514 ,,	15	·016573	·338728	·657008
484 ,,	515 ,,	16	·015608	·344335	·618736
483 ,,	516 ,,	17	·014640	·358975	·580364
482 ,,	517 ,,	18	·013677	·372652	·542197
481 ,,	518 ,,	19	·012726	·385378	·504516
480 ,,	519 ,,	20	·011794	·397172	·467576
479 ,,	520 ,,	21	·010887	·408060	·431609
478 ,,	521 ,,	22	·010008	·418070	·396815
477 ,,	522 ,,	23	·009166	·427236	·363366
476 ,,	523 ,,	24	·008360	·435595	·331407

GROUPS OF			Rank of Groups.	SCALE OF POSSIBILITY. Probability of Drawing each Group. *Table A.*	SCALE OF PRECISION. Sum of the Probabilities, commencing at the most probable Group. *Table B.*	SCALE OF POSSIBILITY. Relative Probability of Drawing each Group. *Table C.*
475 white balls and 524 black			25	·007594	·443189	·301050
474	,,	525 ,,	26	·006871	·450060	·272378
473	,,	526 ,,	27	·006191	·456251	·245451
472	,,	527 ,,	28	·005557	·461809	·220300
471	,,	528 ,,	29	·004968	·466776	·196935
470	,,	529 ,,	30	·004423	·471199	·175343
469	,,	530 ,,	31	·003922	·475122	·155493
468	,,	531 ,,	32	·003464	·478586	·137337
467	,,	532 ,,	33	·003047	·481633	·120816
466	,,	533 ,,	34	·002670	·484304	·105855
465	,,	534 ,,	35	·002330	·486634	·092375
464	,,	535 ,,	36	·002025	·488659	·080290
463	,,	536 ,,	37	·001753	·490412	·069504
462	,,	537 ,,	38	·001512	·491924	·059926
461	,,	538 ,,	39	·001298	·493222	·051461
460	,,	539 ,,	40	·001110	·494332	·044014
459	,,	540 ,,	41	·0009458	·495278	·037493
458	,,	541 ,,	42	·0008024	·496081	·031810
457	,,	542 ,,	43	·0006781	·496759	·026880
456	,,	543 ,,	44	·0005707	·497329	·022623
455	,,	544 ,,	45	·0004784	·497808	·018963
454	,,	545 ,,	46	·0003994	·498207	·015831
453	,,	546 ,,	47	·0003321	·498539	·013164
452	,,	547 ,,	48	·0002750	·498814	·010902
451	,,	548 ,,	49	·0002268	·499041	·008993
450	,,	549 ,,	50	·0001863	·499227	·007387
449	,,	550 ,,	51	·0001525	·499380	·006044
448	,,	551 ,,	52	·0001242	·499504	·004925

S

GROUPS OF	Rank of Groups.	SCALE OF POSSIBILITY. Probability of Drawing each Group. Table A.	SCALE OF PRECISION. Sum of the Probabilities, commencing at the most probable Group. Table B.	SCALE OF POSSIBILITY. Relative Probability of Drawing each Group. Table C.
447 white balls and 552 black	53	·0001008	·499605	·003997
446 „ 553 „	54	·0000815	·499686	·003231
445 „ 554 „	55	·0000656	·499752	·002601
444 „ 555 „	56	·0000526	·499804	·002086
443 „ 556 „	57	·0000421	.499847	·001669
442 „ 557 „	58	·0000334	.499880	·001324
441 „ 558 „	59	·0000265	·499906	·001049
440 „ 559 „	60	·0000209	·499927	·000828
439 „ 560 „	61	·0000164	·499944	·000650
438 „ 561 „	62	·0000128	·499957	·000509
437 „ 562 „	63	·0000100	·499967	·000397
436 „ 563 „	64	·0000077	·499974	·000308
435 „ 564 „	65	·0000060	·499980	·000238
434 „ 565 „	66	·0000046	·499985	·000183
433 „ 566 „	67	·0000035	·499988	.000140
432 „ 567 „	68	·0000027	·4999912	·000107
431 „ 568 „	69	·0000021	·4999933	·000081
430 „ 569 „	70	·0000016	·4999948	·000062
429 „ 570 „	71	·0000012	·4999960	·000047
428 „ 571 „	72	·0000009	·4999969	·000035
427 „ 572 „	73	·0000007	·4999976	·000026
426 „ 573 „	74	·0000005	·4999981	·000020
425 „ 574 „	75	·0000004	·4999984	·000014
424 „ 575 „	76	·0000003	·4999987	·000011
423 „ 576 „	77	·0000002	·4999989	·000008
422 „ 577 „	78	·00000014	·4999990	·000006
421 „ 578 „	79	·00000011	·4999991	·000004
420 „ 579 „	80	·00000004	·4999992	·000003

The table on page 60 and the preceding scale of possibilities only give the numerical coefficients in the development of the binomial. It would then be unnecessary to stop here, were it not very difficult, and even impossible, to calculate these coefficients by ordinary means when the binomial is to be developed to a high power. It would, besides, be useless to calculate 500 coefficients, since 70 or 80 only should be employed.

I will show the method of calculation I have followed; and afterwards how, from a limited number of chances, we can pass to the case of an unlimited number of chances, and embrace the problem in its greatest generality.

We know that the development of the binomial to the power m gives

$$(a + b)^m = a^m + m\, a^{m-1}\, b + \frac{m\,(m-1)}{1.\,2}\, a^{m-2}\, b^2$$

$$+ \ldots + \frac{m\,(m-1)\, \ldots\, (m-n+1)}{1.\,2.\,3\, \ldots\, n}\, a^{m-n}\, b^n.$$

When we consider a and b as the probabilities of two simple events A and B, we also know that

a^m shows the probability that in m trials A will happen m times.

$m\, a^{m-1}\, b$	,,	,,	A	,,	$m-1$ times.
$\dfrac{m\,(m-1)}{1.\,2}\, a^{m-2}\, b^2$	,,	,,	A	,,	$m-2$ times.

$$\frac{m\,(m-1\, \ldots\, (m-n+1)}{1.\,2.\,3\, \ldots\, n}\, a^{m-n}\, b^n \quad \text{,,} \qquad \text{A} \qquad \text{,,} \qquad m-n \text{ times.}$$

When the two probabilities a and b are equal, as I have supposed throughout this note, the development is reduced to the coefficients of the different terms; and we have

$$2^m = 1 + m + \frac{m\,(m-1)}{1.\,2} + \ldots + \frac{m\,(m-1)\,(m-2)\ldots(m-n+1)}{1.\,2.\,3\, \ldots\, n}.$$

The coefficient of each term designates the number of chances which each compound event has; and the total number of these chances is represented by

$$2^m.$$

We see, first, that the total number of chances must be considerable when m is a very large number. Indeed, passing to logarithms, we have

$$\text{Log. } 2^m = m\ 0\cdot 30103:$$

consequently 2^m will be represented by a number which has as many figures as there are units plus 1 in m multiplied by 0.3.

Now this is the course which has been pursued to form the scale of possibility. Let us consider, first, that the general term in the development of the binomial is

$$\frac{m(m-1)(m-2) \dots (m-n+1)}{1.2.3 \quad \dots \quad n},$$

and that the following term would be

$$\frac{m(m-1)(m-2) \dots (m-n+1)(m-n)}{1.2.3 \quad \dots \quad n(n+1)}.$$

Now the ratio of these two quantities, which I shall represent by y_n and y_{n+1}, is

$$y_n : y_{n+1} = 1 : \frac{m-n}{n+1}.$$

To pass to the following term then, it will be sufficient to multiply the term known by $\frac{m-n}{n+1}$, since

$$y_{n+1} = y_n \times \frac{m-n}{n+1}.$$

But I have taken as unity the middle term in the expansion of the binomial,—the term which is the greatest in the hypothesis which we have made. It represents the probability that, in our drawing of 999 balls, there will be 500 white and 499 black, or 499 white and 500 black.

The relative probability of the next event (that is, of drawing 498 white balls and 501 black) is

$$\frac{m-n}{n+1} = \frac{999-500}{500+1} = \frac{499}{501}.$$

The probability of drawing 497 white balls and 502 black will be

$$\frac{499.498}{501.502},$$

and so on. This calculation does not present great difficulties: by following the series of numbers given in the last column of the preceding table, page 256, it has been formed. We may easily deduce from it the numbers in the first column given under the title of *scale of possibility*, Table A.

Designating by $a\ a'\ a''$, &c. each of the numbers calculated in Table C, and taking their sum, we shall have

$$\Sigma a.$$

Then each number of the new column, or Table A, expressing the absolute probability of each of the compound events, will successively become

$$\frac{a}{\Sigma a}, \frac{a'}{\Sigma a}, \frac{a''}{\Sigma a}, \text{ &c.}$$

It is thus that the numbers in the first column were formed. Those in the second column, or Table B, were formed simply by addition in the following manner:—

$$\frac{a}{\Sigma a}, \frac{a + a'}{\Sigma a}, \frac{a + a' + a''}{\Sigma a}, \text{ &c.}$$

The method just indicated is expeditious to calculate; but it may be desired to know the algebraic expression of any term whatever of the development when the number of terms is infinitely great. Let us take again for this purpose the general term

$$\frac{m(m - 1)(m - 2) \dots (m - n + 1)}{1.\,2.\,3 \quad \dots \quad n}.$$

We may write this term under the form

$$\frac{m(m - 1)(m - 2) \dots (m - n + 1)}{1.\,2.\,3 \quad \dots \quad n} \times \frac{(m - n) \dots 3.\,2.\,1}{(m - n) \dots 3.\,2.\,1}.$$

This expression is simplified by the following considerations. We have generally [*]

$$1.\,2.\,3 \dots n = n^n e^{-n} \sqrt{2\pi n} \left(1 + \frac{1}{12n} + \frac{1}{288n^2}\right) + \text{ &c.}$$

π is the ratio of the circumference to the diameter, and e the base of the naperian logarithms. But in the use which we shall make of this formula, we shall always suppose n so great that the second member of this formula can be reduced to its first term. The general term of the *binomial* will thus become

$$y_{n+1} = \frac{m^m e^{-m} \sqrt{2\pi m}}{n^n e^{-n} \sqrt{2\pi n} \times (m - n)^{m-n} e^{-m+n} \sqrt{2\pi (m - n)}}.$$

Such is the formula which shows the number of chances which each event has in its favour. It here refers to the number of chances of the event compounded of A happening $m - n$ times, and B n times.

Let us examine what will be the value of the *maximum* term in our expansion. This term being equally distant from the extremes, we shall have $n = \frac{m}{2}$; and substituting this in the preceding formula, it will become

---

[*] This formula, due to Stierling, is often employed in the calculus of probabilities. See Laplace, *Théorie Analytique des Probabilités*, p. 129; and Poisson, *Recherches sur les Probabilités des Jugements*, p. 175.

$$\text{Mean term} = 2^m \sqrt{\frac{2}{\pi m}}.$$

Now this quantity divided by 2^m, the sum of all the chances, will give the *probability* of the mean term P, which will be

$$P = \sqrt{\frac{2}{\pi m}}.$$

We see that *this probability decreases in the inverse ratio of the square root of m*,—that is to say, that the probability that the two events are the same in number and counterbalance is in the inverse ratio of the square root of the number of observations. The more the number of observations increases, the more the probability of the mean term diminishes.

Comparing the probability of any compound event whatsoever to that of the mean term, we shall have

$$p_{n+1} : P = \frac{m^m e^{-m} \sqrt{2\pi m}}{n^n e^{-n} \sqrt{2\pi n} \times (m-n)^{m-n} e^{-m+n} \sqrt{2\pi(m-n)}} : 2^m \sqrt{\frac{2}{\pi m}}.$$

Reducing, we find

$$p_{n+1} = P \frac{\left(\frac{m}{2}\right)^{m+1}}{n^{n+\frac{1}{2}} \times (m-n)^{m-n+\frac{1}{2}}}.$$

Let us make $m = n + n'$; and let us notice that the numerator multiplied by P is a constant quantity, which we may represent by C. The equation will take this more simple form,—

$$p_{n+1} = \frac{C}{n^{n+\frac{1}{2}} n'^{n'+\frac{1}{2}}}.$$

This last expression shows that the probability of any compound event whatever is inversely proportional to the product of its distances from the two extremes of the development of the binomial : these distances being respectively raised to the powers designated by their rank plus $\frac{1}{2}$.

For every other expansion of the binomial to the power μ we shall have, in like manner,

$$\pi_{\nu+1} = \frac{\Pi\left(\frac{\mu}{2}\right)^{\mu+1}}{\nu^{\nu+\frac{1}{2}} \nu'^{\nu'+\frac{1}{2}}}.$$

Let us put $\mu = ma$, and let us also make $\nu = na$ and $\nu' = n'a$: the ordinate which represents the general term will be raised on the axis of the abscissæ to the point corresponding to that occupied by the ordinate of the first line. Now these substitutions in the formula will give

$$\pi_{\nu+1} = \frac{\Pi \left(\dfrac{ma}{2}\right)^{ma+1}}{(na)^{na+\frac{1}{2}} \cdot (n'a)^{n'a+\frac{1}{2}}}$$

The quantities m and n being very great in comparison with unity, we may write

$$\frac{\pi_{\nu+1}}{\Pi} = \left\{ \frac{\left(\dfrac{ma}{2}\right)^{m+1}}{(na)^{n+\frac{1}{2}} \cdot (n'a)^{n'+\frac{1}{2}}} \right\}^{a} = \left\{ \frac{\left(\dfrac{m}{2}\right)^{m+1}}{n^{n+\frac{1}{2}} \cdot n'^{n'+\frac{1}{2}}} \right\}^{a}.$$

This last quantity is no other than the value of $\dfrac{p_{n+1}}{P}$ raised to the power a. We have then

$$\frac{\pi_{\nu+1}}{\Pi} = \left(\frac{p_{n+1}}{P}\right)^{a}.$$

Because $\mu = ma$, we have again

$$\left(\frac{\pi_{\nu+1}}{\Pi}\right)^{\frac{1}{\mu}} = \left(\frac{p_{n+1}}{P}\right)^{\frac{1}{m}} = \text{a constant.}$$

In Table C, page 256, we have calculated all the appreciable values of

$$\frac{p_{n+1}}{P} \quad \text{or} \quad p_{n+1},$$

taking P as unity. These ratios then agree for any other curve whatever, after they have been raised to the power a, and they correspond respectively to ordinates similarly situated on the axes m and μ.

In the same table we have taken as unity the greatest co-ordinate: there is nothing to prevent us doing the same with another curve; and we have

$$\left(\pi_{\nu+1}\right)^{\frac{1}{\mu}} = \left(p_{n+1}\right)^{\frac{1}{m}}.$$

We will now see how to effect the passage from the finite to the infinite, or the substitution for the consideration of a limited number of chances of an infinitely great number of chances. We have seen that to pass from one ordinate to another we must write

$$y_{n+1} = y_n \frac{m-n}{n+1}.$$

But m is the entire axis of the abscissæ, which we represent by $2m_1$, reckoning from the ordinate at the middle; and taking the abscissæ also from thence, we shall have $n = m_1 - x$, which will give

$$y_{n+1} = y_n \frac{2m_1 - m_1 + x}{m_1 - x + 1} = y_n \frac{m_1 + x}{m_1 - x + 1}.$$

From the ordinate y_{n+1} let us take off that which immediately follows,—

$$y_{n+1} - y_n = y_n \frac{2x - 1}{m_1 - x + 1}.$$

The sign of the second member will depend on whether y_{n+1} is greater or less than y_n. We will here take the sign — .

In considering the difference of the ordinates of the first member, in passing from finite to infinite, as expressing the tangent of the curve, their distance being taken as unity, we shall have

$$\frac{dy}{dx} = -\ y \frac{2x}{m_1 - x},$$

remarking also that m_1 and x are infinite quantities in respect to our unity. But these quantities differ considerably from one another: the ordinates have no appreciable value except in the neighbourhood of the ordinates at the middle, when we suppose the trials infinitely multiplied. We may then in the denominator neglect x without sensible error in presence of m_1: we have definitely

$$\frac{dy}{dx} = -\ \frac{2x}{m_1}\ y.$$

To integrate, we will write

$$\frac{dy}{y} = -\ \frac{2x\,dx}{m_1},$$

whence

$$\text{Log. } y = -\ \frac{x^2}{m_1} + \text{a constant.}$$

For $x = 0$ we have the *maximum* ordinate at the middle of the curve: I will designate it by Y. We have then, returning to numbers,

$$y = Y e^{-\frac{x^2}{m_1}}.$$

Such is the equation of the curve of possibility. There exists a ratio which may be assigned between the sum of all the ordinates of the curve, or its surface, and the ordinate at the middle. This remarkable ratio is $\sqrt{\pi m}$; so that we have

$$\text{The surface or S} = Y \sqrt{\pi m}.$$

π is the ratio of the circumference to the diameter.

When we are considering only one curve, it is a matter of indifference what unit we take for the abscissæ and the ordinates. If we take $\dfrac{1}{\sqrt{m_1}}$ as the unit of the abscissæ, the first equation becomes

$$y = Y e^{-x^2}.$$

By this we obtain, as the surface of the curve,

$$S = Y \sqrt{\pi}.$$

But this surface represents the sum of all the probabilities, which should be equal to 1, the symbol of certainty. We have then

$$S = 1 = Y \sqrt{\pi}, \text{ and } Y = \frac{1}{\sqrt{\pi}};$$

or again, as the equation of the curve,

$$y = \frac{1}{\sqrt{\pi}} e^{-x^2}.$$

The surface of the curve contained between the mean ordinate and another ordinate, the abscissa of which is n, may be considered as expressing the probability of the errors comprised from 0 to x. We should then have

$$\text{The surface} = \int_x^0 y\,dx = \frac{1}{\sqrt{\pi}} \int_x^0 e^{-x^2}\,dx.$$

We might integrate by using the known development *

$$e^x = 1 + z + \frac{z^2}{1.\,2} + \frac{z^3}{1.\,2.\,3} + \&c.;$$

but there are more expeditious methods.

M. Cournot, in his *Exposition de la Théorie des Chances et des Probabilités*, has calculated by means of the preceding formula a sufficiently extended table, which we have here reproduced to compare it with our own.† The values have been deduced from those of the function $\int_t^\infty e^{-t^2}\,dt$ given by Kramp in *L'Analyse des Réfractions Astronomiques*.

We can now judge if our table of precision, calculated on the hypothesis of a thousand different events, differs much from that which would be obtained in adopting the case of nature, in which the probability of the expected event may pass through every possible gradation.

---

* M. Hagen has calculated a small table after this method: it is given on page 50 of his work *Grundzüge der Wahrscheinlichkeits-Rechnung*, Berlin, 1837.

† This work was published in 1843, by Hachette, at Paris. I was not aware of it, nor of the preceding, until after my tables were calculated.

Let us first examine how the numbers given in our table correspond with those in the table calculated by M. Cournot. It will, however, only be necessary to double them for the purpose of comparison. Our table in fact gives the sum of the probabilities for one side of the axis of symmetry only, and that of M. Cournot for both sides at once. Not to extend the table too much, let us confine ourselves to the ten first numbers of our Table B, doubling them, then those which follow at the distance of ten ranks. The first column indicates this rank, and the second the corresponding rank in the French table.

| RANK OF PROBABILITIES. | | VALUE OF THE PROBABILITY. | RANK OF PROBABILITIES. | | VALUE OF THE PROBABILITY. |
Table B.	Table of M. Cournot.		Table B.	Table of M. Cournot.	
1	4·5	0·050450	9	40·3	0·430958
2	9·0	0·100698	10	44·8	0·473096
3	13·5	0·150546	20	89·5	0·794344
4	18·0	0·199800	30	134·5	0·942398
5	22·4	0·248272	40	179·0	0·988664
6	26·9	0·295784	50	224·0	0·998454
7	31·4	0·342170	60	269·0	0·999854
8	36·8	0·387274	67	299·0	0·999976

The probability placed in the 67th rank in our table will be found in the 299th in that of M. Cournot. Thus the result $\frac{299}{67}$ nearly equals 4·5. This is also the result which would be found for the respective places of all the other probabilities in the two calculated tables; which thus presents as satisfactory an agreement as could be desired. We see, moreover, without having recourse to means of integration, that we should have reproduced M. Cournot's table in calculating the development of the binomial, as we have done to the power 1,000 $(\frac{299}{67})^2$ or to the power of about 20,000.

TABLE

Of the values of the functions $P = \dfrac{2}{\sqrt{\pi}} \displaystyle\int_0^t e^{-t^2}\, dt,$

Calculated after that of the values of the function $\displaystyle\int_t^\infty e^{-t^2}\, dt$, given by Kramp in the *Analyse des Réfractions Astronomiques*, Strasbourg, year VII.

t	P	t	P	t	P	t	P
0·00	0·000 00	0·23	0·255 02	0·46	0·484 66	0·69	0·670 84
0·01	0·011 28	0·24	0·265 70	0·47	0·493 74	0·70	0·677 80
0·02	0·022 57	0·25	0·276 32	0·48	0·502 75	0·71	0·684 67
0·03	0·033 84	0·26	0·286 90	0·49	0·511 67	0·72	0·691 43
0·04	0·045 11	0·27	0·297 42	0·50	0·520 50	0·73	0·698 10
0·05	0·056 37	0·28	0·307 88	0·51	0·529 24	0·74	0·704 68
0·06	0·067 62	0·29	0·318 28	0·52	0·537 90	0·75	0·711 16
0·07	0·078 86	0·30	0·328 63	0·53	0·546 46	0·76	0·717 54
0·08	0·090 08	0·31	0·338 92	0·54	0·554 94	0·77	0·723 82
0·09	0·101 28	0·32	0·349 13	0·55	0·563 32	0·78	0·730 01
0·10	0·112 46	0·33	0·359 28	0·56	0·571 62	0·79	0·736 10
0·11	0·123 62	0·34	0·369 36	0·57	0·579 82	0·80	0·742 10
0·12	0·134 76	0·35	0·379 38	0·58	0·587 92	0·81	0·748 00
0·13	0·145 87	0·36	0·389 33	0·59	0·595 94	0·82	0·753 81
0·14	0·156 95	0·37	0·399 21	0·60	0·603 86	0·83	0·759 52
0·15	0·168 00	0·38	0·409 01	0·61	0·611 68	0·84	0·765 14
0·16	0·179 01	0·39	0·418 74	0·62	0·619 41	0·85	0·770 67
0·17	0·189 99	0·40	0·428 39	0·63	0·627 05	0·86	0·776 10
0·18	0·200 94	0·41	0·437 97	0·64	0·634 59	0·87	0·781 44
0·19	0·211 84	0·42	0·447 47	0·65	0·642 03	0·88	0·786 69
0·20	0·222 70	0·43	0·456 89	0·66	0·649 38	0·89	0·791 84
0·21	0·233 51	0·44	0·466 23	0·67	0·656 63	0·90	0·796 91
0·22	0·244 30	0·45	0·475 48	0·68	0·663 78	0·91	0·801 88

t	P	t	P	t	P	t	P
0·92	0·806 77	1·24	0·920 50	1·56	0·972 628	1·88	0·992 156
0·93	0·811 56	1·25	0·922 90	1·57	0·973 602	1·89	0·992 479
0·94	0·816 27	1·26	0·925 24	1·58	0·974 546	1·90	0·992 790
0·95	0·820 89	1·27	0·927 51	1·59	0·975 461	1·91	0·993 089
0·96	0·825 42	1·28	0·929 73	1·60	0·976 348	1·92	0·993 378
0·97	0·829 87	1·29	0·931 90	1·61	0·977 206	1·93	0·993 655
0·98	0·834 23	1·30	0·934 01	1·62	0·978 038	1·94	0·993 922
0·99	0·838 51	1·31	0·936 06	1·63	0·978 842	1·95	0·994 179
1·00	0·842 70	1·32	0·938 06	1·64	0·979 621	1·96	0·994 426
1·01	0·846 81	1·33	0·940 01	1·65	0·980 375	1·97	0·994 663
1·02	0·850 84	1·34	0·941 91	1·66	0·981 104	1·98	0·994 891
1·03	0·854 78	1·35	0·943 76	1·67	0·981 810	1·99	0·995 111
1·04	0·858 65	1·36	0·945 56	1·68	0·982 492	2·00	0·995 3223
1·05	0·862 44	1·37	0·947 31	1·69	0·983 152	2·01	0·995 5248
1·06	0·866 14	1·38	0·949 02	1·70	0·983 790	2·02	0·995 7194
1·07	0·869 77	1·39	0·950 67	1·71	0·984 406	2·03	0·995 9064
1·08	0·873 33	1·40	0·952 28	1·72	0·985 002	2·04	0·996 0859
1·09	0·876 80	1·41	0·953 85	1·73	0·985 578	2·05	0·996 2580
1·10	0·880 20	1·42	0·955 38	1·74	0·986 134	2·06	0·996 4235
1·11	0·883 53	1·43	0·956 86	1·75	0·986 671	2·07	0·996 5821
1·12	0·886 79	1·44	0·958 30	1·76	0·987 190	2·08	0·996 7344
1·13	0·889 97	1·45	0·959 69	1·77	0·987 690	2·09	0·996 8805
1·14	0·893 08	1·46	0·961 05	1·78	0·988 174	2·10	0·997 0206
1·15	0·896 12	1·47	0·962 37	1·79	0·988 640	2·11	0·997 1548
1·16	0·899 10	1·48	0·963 65	1·80	0·989 090	2·12	0·997 2836
1·17	0·902 00	1·49	0·964 90	1·81	0·989 524	2·13	0·997 4070
1·18	0·904 84	1·50	0·966 105	1·82	0·989 943	2·14	0·997 5253
1·19	0·907 61	1·51	0·967 276	1·83	0·990 346	2·15	0·997 6386
1·20	0·910 31	1·52	0·968 413	1·84	0·990 735	2·16	0·997 7471
1·21	0·912 96	1·53	0·969 516	1·85	0·991 111	2·17	0·997 8511
1·22	0·915 53	1·54	0·970 585	1·86	0·991 472	2·18	0·997 9507
1·23	0·918 05	1·55	0·971 622	1·87	0·991 820	2·19	0·998 0459

t	P	t	P	t	P
2·20	0·998 1371	2·47	0·999 '5226	2·74	0·999 893 35
2·21	0·998 2244	2·48	0·999 5472	2·75	0·999 899 38
2·22	0·998 3079	2·49	0·999 5707	2·76	0·999 905 08
2·23	0·998 3878	2·50	0·999 593 05	2·77	0·999 910 48
2·24	0·998 4642	2·51	0·999 614 29	2·78	0·999 915 59
2·25	0·998 5373	2·52	0·999 634 50	2·79	0·999 920 42
2·26	0·998 6071	2·53	0·999 653 71	2·80	0·999 924 99
2·27	0·998 6739	2·54	0·999 671 98	2·81	0·999 929 31
2·28	0·998 7375	2·55	0·999 689 34	2·82	0·999 933 39
2·29	0·998 7986	2·56	0·999 705 84	2·83	0·999 937 25
2·30	0·998 8568	2·57	0·999 721 51	2·84	0·999 940 90
2·31	0·998 9124	2·58	0·999 736 40	2·85	0·999 944 34
2.32	0·998 9655	2·59	0·999 750 54	2·86	0·999 947 60
2·33	0·999 0162	2·60	0·999 763 96	2·87	0·999 950 67
2·34	0·999 0646	2·61	0·999 776 71	2·88	0·999 953 58
2·35	0·999 1107	2·62	0·999 788 81	2·89	0·999 956 32
2·36	0·999 1548	2·63	0·999 800 29	2·90	0·999 958 90
2·37	0·999 1968	2·64	0·999 811 18	2·91	0·999 961 34
2·38	0·999 2369	2·65	0·999 821 52	2·92	0·999 963 65
2·39	0·999 2751	2·66	0·999 831 31	2·93	0·999 965 82
2·40	0·999 3115	2·67	0·999 840 60	2·94	0·999 967 86
2·41	0·999 3462	2·68	0·999 849 41	2·95	0·999 969 80
2·42	0·999 3793	2·69	0·999 857 76	2·96	0·999 971 62
2·43	0·999 4108	2·70	0·999 865 67	2·97	0·999 973 33
2·44	0·999 4408	2·71	0·999 873 16	2·98	0·999 974 95
2·45	0·999 4694	2·72	0·999 880 26	2·99	0·999 976 47
2·46	0·999 4966	2·73	0·999 886 98		

$t = 3·00$ $\quad$ P $= 0·999$ 977 909 3 . . . .

$4·00$ $\qquad$ 0·999 999 984 582 8 . . . .

$5·00$ $\qquad$ 0·999 999 999 998 432 53 . . . .

LETTER XVII., page 78.

"Compared with the theory of balances or of levers in mechanics, the Theory of Means offers curious analogies."

There are different lengths the mean of which we wish to ascertain. The length A is repeated p times; the length A', p' times; and so on. If M be the mean of all the lengths, we should have

$$M(p + p' + p'' + p''' + \&c.) = Ap + A'p' + A''p'' + A'''p'''. + \&c.$$

But we may take

$$A = M - a, \ A' = M - a', \ A'' = M - a'', \&c.;$$

a, a', a'', &c. being the distances from the extremity of the mean M, while A, A', A'', &c. are the distances from a common origin. This substitution gives

$$Ap + A'p' + A''p'' + A'''p''' + \&c. = M(p + p' + p'' + p''' + \&c.)$$
$$- (ap + a'p' + a''p'' + \&c.)$$

By the first equation we shall have

$$ap + a'p' + a''p'' + a'''p''' + \&c. = 0.$$

Under a mechanical view, the statical moments $ap, a'p', a''p''$ form an equilibrium around the extreme point of the mean M, which corresponds with the centre of gravity.

The first equation shows, moreover, that for all the levers A, A', A'' A''', &c., loaded with weights proportional to the numbers p, p', p'', p''', and placed at the extremity opposite to the common origin, we may substitute a single lever M loaded with the sum of all the weights p, p', p'', &c.

LETTER XVIII., page 82.

"It is the greatness of the probable error which will henceforth serve us for a *modulus of precision*."

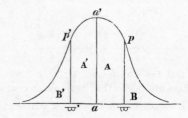

The surface of the curve of precision is represented by I, as forming the sum of all the probabilities relating to the same event. We have seen that that represented by the abscissa, which corresponds to the semi-surface of the curve, is called *probable error*. Thus, on the two sides of the maximum ordinate aa', and at equal distances $a\omega\ a\omega'$, we must conceive two equal ordinates ωp and $\omega'p'$, between which is comprised a portion $\omega'\omega pa'p'$ of the surface equivalent to those found beyond the same ordinates to the extreme limits of the curve. In the first fall the smallest errors; it may be represented by $A + A'$, and the other by $B + B'$, on account of the symmetry of the curve, $A = A'$ and $B = B'$. Then we may take the two portions of surface A and B in different ratios; thus, for example, B may be to A as 1 is to 1, 2, 4, 8, &c. The portion of surface A which corresponds to the least errors will be successively $\frac{1}{2}, \frac{2}{3}, \frac{4}{5}, \frac{8}{9}, \frac{16}{17}$, &c., of the semi-surface S of of the curve,—that is to say, the ordinate $p\omega$ differs successively from aa'. Let us designate by P the ratio of A to the demi-surface of the curve or to $\frac{1}{2}$, and by π the abscissa $a\omega$, or the error which corresponds to it: we may form the following table.*

Ratio between the Portions of Surface B and A.	$P = \dfrac{A}{2\,(A + B)}$ or Portion of the Demi-surface of the Curve S.	π the Error, or the Rank in the Scale of Precision.	Magnitude of the Error π in relation to the probable Error ω.
1 : 1	0·2500	10·67 = ω	1·00 × ω
1 : 2	0·3333	15·4	1·43 × ω
1 : 4	0·4000	20·3	1·90 × ω
1 : 8	0·4444	25·2	2·36 × ω
1 : 16	0·4706	29·9	2·80 × ω
1 : 32	0·4848	34·2	3·21 × ω
1 : 64	0·4923	31·3	3·59 × ω
1 : 128	0·4961*	42·1	3·95 × ω
1 : 256	0·4981	45·7	4·38 × ω
1 : 512	0·4990	49·0	4·59 × ω
1 : 1024	0·4995	52·0	4·88 × ω

It will be seen that the chances are 1 to 1 that there will be as many errors greater as there will be less than the probable error ω. The chances

* This table is borrowed from the German work of M. Hagen, page 57.

are 1 to 2 that the error will be 1·43 of the probable error; 1 to 4 that the error will be 1·90 of the probable error; and 1 to 1,024 that the error will be 4·88 times the probable error.

From the preceding table the following is deduced by interpolation :—

THE CHANCES ARE	THAT THE ERROR WILL NOT EXCEED.
1 to 1	1 ω = probable error.
4·6383 to 1	2 ω
22·239 to 1	3 ω
158·59 to 1	4 ω
1384·0 to 1	5 ω
20000 to 1	6 ω

We may then risk 20,000 to 1 that the error will not exceed six times the probable error.

The ordinate $p\,\omega$, which corresponds to the probable error, being compared with the maximum ordinate of the curve, equals 0·84536 according to Table C, indicating the relative probability of each drawing. Designating the first ordinate by p, and the maximum ordinate by a, we have then

$$p = 0\cdot84536\,a.$$

The probable errors + ω and — ω correspond to the two equal values of p.

In a series of observations, as we have seen, the errors are distributed in proportion to the lengths of the ordinates of the curve. When the method of determining the probable error is once known, we can easily discover by the preceding table the probability of an error of a given magnitude being made.

LETTER XVIII., page 80.

Let x, x', x'', x''', &c., be the results of observations equally deserving of confidence,—their arithmetical mean will be the most probable value of the number sought.* It is

$$X = \frac{x + x' + x'' + x'' + \&\text{c.}}{N} = \frac{1}{N}\,\Sigma\,x.$$

* This note is partially extracted from the Dictionary of Gehler (article WAHRSCHEIN-LICHKEITS-RECHNUNG), and from the work of Hagen.

N is the number of observations; Σx is their sum. Let ϵ be the difference between our mean X and a particular value x, we shall have

$$\epsilon = X - x$$
$$\epsilon' = X - x'$$
$$\epsilon'' = X - x''$$
$$\cdot \qquad \cdot$$
$$\cdot \qquad \cdot$$
$$\cdot \qquad \cdot$$

The quantities ϵ, ϵ', ϵ'' may be considered as the errors of particular observations. Let us write

$$\Sigma \epsilon^2 = \epsilon^2 + \epsilon'^2 + \epsilon''^2 + \epsilon'''^2 + \&c.$$

The quantity

$$P = \frac{N^2}{2\Sigma \epsilon^2}$$

is the weight *(pondus)* of the value X,—that is to say, the weight of the value of the mean of all the quantities x, x', x'', x'''. It will be easily seen that P will be greater as the number N of observations are more considerable, and as the errors ϵ, ϵ', ϵ'' are less.

We shall remark that the quantities ϵ, ϵ', ϵ'' are the abscissæ, and the quantities x, x', x'' the ordinates of our curve of possibility.

When the chances of error in excess are equal to the chances of error in defect, the positive and negative errors place themselves symmetrically on each side of the mean;

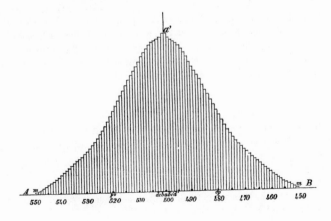

and we have

$$+ \epsilon, - \epsilon', + \epsilon'_1 - \epsilon'_1 + \epsilon''_1 - \epsilon'', \&c.$$

T

The squares of the errors give the sum

$$2\varepsilon^2 + 2\varepsilon'^2 + 2\varepsilon''^2 + \&\text{c.} = 2\Sigma\varepsilon^2.$$

Let Q^2 represent the arithmetical mean of the squares of all the errors, we shall have

$$Q^2 = \frac{2\Sigma\varepsilon^2}{N}.$$

This quantity Q^2 is of great importance. In making it enter into the expression of the weight of the mean, we have

$$P = \frac{N}{Q^2}.$$

According to its property of mean, it follows that $Q^2 = \frac{1}{2}$; whence $Q = 0\cdot70711$ in parts of unity of which $\omega = 0\cdot4769365$. Thus

$$\omega = \frac{0\cdot4769365}{0\cdot70711} Q = 0\cdot67450.\ Q,$$

and reciprocally

$$Q = 1\cdot4826\ \omega.$$

ω is the extent of the probable error or the abscissa of the curve of possibility for the surface $\frac{1}{4}$; its rank in our table of precision is nearly $10\cdot66$, and consequently the rank of Q in this same table is $15\cdot805$,—that is to say, it is necessary to keep the limit within this range, so that the surface of the curve of possibility may be $0\cdot70711$, and represent the square root of the arithmetical mean of the square of all the errors.

It must be remarked that ω only expresses the abstract number $0\cdot4769363$; when we pass to the application the quantity ω becomes a concrete number. Let E be the probable error produced in the determination of X; it is, as we have seen, the abscissa of the curve of possibility which corresponds to the portion of surface equivalent to that placed on the remainder of the axis. We have for that probable error

$$E = \frac{\omega}{\sqrt{P}}.$$

If the weight P be unity, the probable error E will be constant, and the value of it will be ω. If a be taken as constant, E will be reciprocally proportional to the square root of the weight of the mean.

Thus the relative calculations concerning the right ascension of the Polar Star have shown us that half of the observations fall between the limits $+ \omega$ and $- \omega = (10\cdot66)$, and that the limits embrace $1^s\cdot61$. Then, on the hypothesis that $P = 1$, we have

$$E = \omega = 0^s\cdot805.$$

The probable error is comprised in the limit $0^s\cdot805$.

Having E, we can deduce the value of the weight P by the formula

$$P = \frac{\omega^2}{E^2} .$$

Equal use is made of the *mean error*, which is the sum of the products of each particular error by its probability. This mean error ϕ, which is produced in determining X, is expressed

$$\phi = \frac{1}{2 \sqrt{\pi P}} = \frac{0.282095}{\sqrt{P}};$$

π being the relation of the circumference to the diameter 3·14159. The two errors E and ϕ are connected thus by the relations

$$\phi = 0.59147\,E; \text{ and } E = 1.69069\,\phi.$$

The quantities ϕ and E relate to the mean X. After the preceding relations, it is evident that the mean error in the calculation of the right ascension of the Polar Star will give

$$\phi = 0.59147 \times 1^{s}\!\cdot\!61\,\phi.$$

For our table, when use is made of the mean error, it is necessary to write

$$\phi = 0.59147 \times 10.66 = 6.30.$$

The mean error corresponds thus to the rank 6·30 in our table of precision.

The *probable error of each individual observation* may also be considered as we have done for the general mean; and then we have

$$e = \omega \sqrt{\frac{N}{P}}; \text{ or } e = E \sqrt{N}.$$

The particular error compared with the probable is larger as N is greater; and the relation of the two errors is as the square root of the number of the observations.

We may then write, in making use of the weight of the arithmetical mean of the squares of all the errors,

$$e = \omega \times Q.$$

The real error may be more or less great for each particular observation, nevertheless the variations are confined within the limits determined by the calculation. The *limits of the real error* of each particular observation are

$$e \left(1 + \frac{0.4769363}{\sqrt{N}}\right), \text{ or } E \left(0.4769363 + \sqrt{N}\right).$$

We may risk 1 to 1 that the true error e falls within the limits

$$E \left(0.4769363 + \sqrt{N}\right), \text{ and } E \left(0.4769363 - \sqrt{N}\right).$$

LETTER XX., page 92.

MEASURE OF THE CIRCUMFERENCES OF THE CHESTS OF SCOTCH SOLDIERS.							
Measures of the Chest.	Number of Men.	Proportional Number.	Probability according to the Observation.	Rank in the Table.	Rank according to Calculation.	Probability according to the Table.	Number of Observations calculated.
INCHES.							
33	3	5	0·5000			0·5000	7
34	18	31	0·4995	52	50	0·4993	29
35	81	141	0·4964	42·5	42·5	0·4964	110
36	185	322	0·4823	33·5	34·5	0·4854	323
37	420	732	0·4501	26·0	26·5	0·4531	732
38	749	1305	0·3769	18·0	18·5	0·3799	1333
39	1073	1867	0·2464	10·5	10·5	0·2466	1838
			0·0597	2·5	2·5	0·0628	
40	1079	1882	0·1285	5·5	5·5	0·1359	1987
41	934	1628	0·2913	13	13·5	0·3034	1675
42	658	1148	0·4061	21	21·5	0·4130	1096
43	370	645	0·4706	30	29·5	0·4690	560
44	92	160	0·4866	35	37·5	0·4911	221
45	50	87	0·4953	41	45·5	0·4980	69
46	21	38	0·4991	49·5	53·5	0·4996	16
47	4	7	0·4998	56	61·8	0·4999	3
48	1	2	0·5000			0·5000	1
	5738	1·0000					1·0000

The preceding table is formed in exactly the same manner as that concerning the measurement of the right ascension of the Polar Star. Consequently I may dispense with entering into new details on the indications of each column.

I will remark that the groups of observations proceed by the differences of 1 English inch, and that the numbers which correspond to them in the calculated table present themselves at 8 ranks of distance. For 21 ranks we shall have then proportionally 2·625 inches, or 66·674^{mm} (millimètres). The probable variation is then about 1·31 inches, or 33·34^{mm}. This variation may appear great: it moreover depends less upon the precision of the measurements, which appear correct, than upon the nature of the elements measured, which were very different.

LETTER XXI., page 95.

HEIGHT OF FRENCH CONSCRIPTS.						
Measures of Height.	Number of Men.	Probability according to the Observation.	Rank in the Table.	Rank according to the Calculation.	Probability according to the Table.	Number of Men calculated.
INCHES. INCHES. 52·244 to 53·307				62·2	0·49996	14
53·307 „ 54·370				56·4	0·49982	47
54·370 „ 55·433				50·6	0·49935	164
55·433 „ 56·496				44·8	0·49771	449
56·496 „ 57·559	28620	0·50000	...	39·0	0·49322	1105
57·559 „ 58·622				33·2	0·48217	2370
58·622 „ 59·685				27·4	0·45847	4440
59·685 „ 60·758				21·6	0·41407	7285
60·758 „ 61·821				15·8	0·34122	10467
61·821 „ 62·884	11580	0·21380	9·	10·0	0·23655	13182
62·884 „ 63·947	13990	0·09800	4·	4·2	0·10473	14502
		0·04190	1·6	1·6	0·04029	
63·947 „ 65·010	14410	0·18600	7·5	7·4	0·18011	13982
56·010 „ 66·073	11410	0·30010	13·5	13·2	0·29814	11803
66·073 „ 67·136	8780	0·38790	19·	19·0	0·38539	8725
67·136 „ 68·199	5530	0·44320	25·	24·8	0·44166	5527
68·199 „ 69·262	3190	0·47510	31·	30·6	0·47355	3187
69·262 „ 70·325				36·4	0·48936	1581
70·325 „ 71·388				42·2	0·49621	685
71·388 „ 72·451	2490	0·50000	...	48·0	0·49881	260
72·451 „ 73·514				53·8	0·49967	86
73·514 „ 74·577				59·6	0·49993	26
74·577 „ 75·640				65·4	0·49998	5
Above 75·640	...	...	...	71·2	0·50000	2
	100000					

In the preceding example, the series of observations was not continued. For the height, all the individuals under 61·821 inches, as also all those above 68·199 inches, have been grouped together. Nevertheless, the seven intermediate groups have been sufficient to establish the ranks which they should occupy in our table. The distance of the groups has been found to be 5·8 units of our scale; and that distance for the heights is equivalent to 1·063 inch. For 21 ranks or units of the scale we should have 3·857 inches; consequently the probable difference would be little less than 1·928 inch. If it is observed that the mean height is about 63·947 inches, we shall also see that the probable difference forms $\frac{1}{33}$ of the quantity measured.

Letter XXI., page 97.

"The number rejected for deficiency in height is much exaggerated. Not only can we prove this, but we can determine the extent of the fraud."

If we compare the numbers observed with the numbers already calculated, we shall be able to give the following table:—

Height of Men.	Number of Men.		Differences of Results.
	Measured.	Calculated.	
INCHES.			
Under 61·821	28620	26345	+ 2275
61·821 to 62·884	11580	13182	− 1602
62·884 „ 63·947	13990	14502	− 512
63·947 „ 65·010	14410	13982	+ 428
65·010 „ 66·073	11410	11803	− 393
66·073 „ 67·136	8780	8725	+ 55
67·136 „ 68·199	5530	5527	+ 3
68·199 „ 69·262	3190	3187	+ 3
Above 69·262	2490	2645	− 155

If we regard the magnitudes of the numbers, it will be observed that the only remarkable difference between the numbers calculated and observed is in the smallest heights. The calculation shows that 2,275 men have been rejected for want of height, which appears ought not to have been the case according to the law of continuity. This number is found wanting in the two extreme categories, which are too small, the one by 1,602 men, and the other by 512 men, giving in the total 2,114, a number little differing from the preceding.

LETTER XXII. (*postscript*), page 104.

" I have been able to measure the different dimensions of the body of the celebrated dwarf whom the United States have offered to the admiration of Europe."

MEASUREMENTS TAKEN OF CHARLES S. STRATTON, SURNAMED
GENERAL TOM THUMB.

PARTS OF THE BODY.	Tom Thumb.	An Infant of 1 to 3 Years.	An Infant of 1 to 3 Years reduced 0·11.
	INCHES.	INCHES.	INCHES.
Height in boots, the soles of which were ·39 of an inch in thickness	27·95	31·41	27·95
Length of extended arms	25·98	30·86	27·47
Height of head	6·03	6·89	6·14
Circumference of head by the frontal sinus . .	17·43	19·37	17·24
From the top of the head to the clavacles . .	6·81	7·43	6·61
Distance of the shoulders between the processes of the shoulder-blades	7·95	7·68	6·85
Circumference of the shoulders between the processes of the shoulder-blades	19·69	18·97	16·83
Circumference of the hips	18·81	19·77	17·59
Length of the arm from the process of the shoulder-blade to the extremity of the hand . . .	9·64	12·55	11·17
Length of the hand	2·95	3·74	3·34
Ditto of the naked foot	4·13	4·89	4·37
Breadth of the hand	1·73	1·84	1·69
Ditto of the naked foot	1·65	2·03	1·80
Height of the leg to the middle of the knee-cap .	6·89	7·60	6·81
From the fork to the ground	10·43	11·21	10·00
From the trochanter to the ground . . .	11·91	13·07	11·81
Circumference of the calf of the leg . . .	6·18	6·69	5·99
Length of the ear	1·84	1·97	1·77

Charles S. Stratton, surnamed General Tom Thumb, was born, according to his memoir, at Bridgeport, Connecticut, U.S., on the 11th January 1832. He was in his fourteenth year when I had the opportunity of seeing him. According to the said memoir, the General at the time of his birth weighed $9\frac{1}{2}$ lbs., or 4 kilogrammes 313 grammes. No accident preceded or followed his birth. Nothing extraordinary was discovered in him until he attained the age of seven months. He then weighed 15 lbs. (English); and it was about this time that he was perceived to cease growing. His parents are of the ordinary stature, and have had two other children, who present no anomaly in their development.

I am indebted to the kindness of M. Clemson, Chargé d'Affaires of the United States at Brussels, for the opportunity of measuring the principal dimensions of the Bridgeport Dwarf. They have been taken with the greatest care, and will be found arranged in the preceding table. I have compared them with the dimensions to which they appear to me to be the most analogous in the tables I have constructed for the different ages, and which I extract from a work *On the Proportions of Man* soon to be published.

The proportions of Tom Thumb, on the 9th July 1845, were nearly those of a child from 1 to 3 years of age reduced 11 per cent. of their value, which may correspond to the dimensions of an ordinary child from 14 to 15 months old.

Letter XXIV., page 110.

" These groups would not be composed, at hazard, of numbers more or less great, although they arise from awkwardness or from defect in sight."

I will endeavour to make it more intelligible, by repeated proofs, how accidental errors correct themselves. I will admit that, with all the precautions I can take for obtaining great exactness, I am necessarily exposed to purely accidental causes of error concerning the truth. For simplification, I will suppose that the causes are only ten in number, although their number may generally be considered as unlimited. I will moreover suppose each of them to give rise to a similar error in the measurement—to an error of $\cdot005^{mm}$, either more or less. Thus the ten causes of error to which I am exposed may concur in the same direction either to give too great or too small a measure. In the first case I shall obtain the greatest possible, and in the second case the smallest possible error. But the probability of the ten causes of error acting at the same time in the same direction will be very slight, as we may easily assure ourselves.

Each of the accidental causes, according to the definition which I have before given, has the same probability of acting in one direction as the other. This probability is then $\frac{1}{2}$. But the probability of the ten errors occurring at the same time in the same direction, and being thus all positive or all negative, belongs to an event composed of ten simple events. It is necessary then to multiply together the ten fractions $\frac{1}{2}$ which express the probabilities of these ten simple events; and the result will be $\frac{1}{1024}$.

We should remark that, by the same kind of calculation, each of the two greatest possible errors only offers a single chance in its favour on a total number of 1024 chances. The 1022 remaining chances belong to nine other possible events.

Ten of these chances are for nine of the ten causes acting positively, and the tenth producing an error in the contrary direction. In this case the nine first causes will produce an error of $+$ 0·45^{mm}; which will be partly counteracted by the single cause of error which shall have acted in the contrary direction, and which would produce an error of 0·05^{mm}. Taking into account the ten causes of error, the real inaccuracy will only be 0·40^{mm}. This error will be the same as that produced by the ten accidental causes, supposing the single one to act positively and the nine negatively. This event will also have ten chances in its favour.

To explain the preceding, we must conceive that the action of nine causes in the same direction, and of a single cause in the contrary direction, may arise in ten different ways; whilst the concurrence of ten causes at the same time, in the same direction, can only be produced in one way. In fact, each of the ten causes, taken singly, may act in a determined direction, whilst the nine others act in a contrary direction. There would then be really ten kinds of possible events, or rather ten distinct chances, which will lead to the same result,—that is to say, a single negative error of 0·05^{mm}, and a positive error of nine times this amount.

The number of possible chances is still greater when it is supposed that two of the accidental causes may act in a determined direction, whilst the eight others act in another.

In general the number of chances is so much the greater in proportion as the modes of action of the causes tend to equality; and the most probable event is that in which five of the ten causes act in one direction, and the other five in the opposite.

The following table will better illustrate what has just been said. By the side of all the possible combinations which can be attributed to the causes in their mode of action will be found the errors to which these causes give rise, and the number of chances which each combination has in its favour.

The Ten Accidental Causes can give rise to				Total Error.	Number of Chances.
10 times the error + 0·05 mm;		0 times the error − 0·05 mm		+ 0·5 mm	1
9 "	"	1 "	"	+ 0·4	10
8 "	"	2 "	"	+ 0·3	45
7 "	"	3 "	"	+ 0·2	120
6 "	"	4 "	"	+ 0·1	210
5 "	"	5 "	"	0·0	252
4 "	"	6 "	"	− 0·1	210
3 "	"	7 "	"	− 0·2	120
2 "	"	8 "	"	− 0·3	45
1 "	"	9 "	"	− 0·4	10
0 "	"	10 "	"	− 0·5	1

LETTER XXVI., page 120.

"The table which accompanies this letter may make this more sensible."

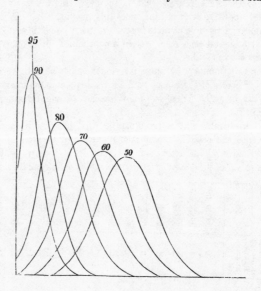

TABLE showing the Probabilities of drawing 16 Balls at a Time from an Urn containing an infinite Number of Black and White Balls in different Proportions.

NUMBER OF BLACK BALLS DRAWN.	OF EVERY 100 BALLS WHICH THE URN CONTAINS THE NUMBER OF WHITE BALLS IS										NUMBER OF BLACK BALLS DRAWN.
	50	55	60	65	70	75	80	85	90	95	
0	0·000 02	0·000 07	0·000 28	0·001 02	0·003 32	0·010 02	0·028 15	0·074 92	0·185 29	0·440 11	0
1	0·000 25	0·000 92	0·003 01	0·008 75	0·022 79	0·053 45	0·112 59	0·202 52	0·329 43	0·370 63	1
2	0·001 83	0·005 63	0·015 05	0·035 32	0·073 25	0·133 63	0·211 10	0·279 98	0·274 52	0·146 31	2
3	0·008 54	0·021 51	0·046 81	0·088 76	0·146 50	0·207 88	0·246 30	0·230 57	0·142 35	0·035 94	3
4	0·027 77	0·057 19	0·101 42	0·155 34	0·204 04	0·225 21	0·200 11	0·132 24	0·051 40	0·006 15	4
5	0·066 65	0·112 26	0·162 27	0·200 76	0·209 88	0·180 16	0·120 07	0·056 01	0·013 71	0·000 78	5
6	0·122 19	0·168 43	0·198 33	0·198 18	0·164 91	0·110 09	0·055 03	0·018 12	0·002 79	0·000 07	6
7	0·174 56	0·196 87	0·188 89	0·152 46	0·100 97	0·052 43	0·019 65	0·004 57	0·000 44	0·000 01	7
8	0·196 38	0·181 21	0·141 66	0·092 36	0·048 68	0·019 66	0·005 53	0·000 91	0·000 06	0·000 00	8
9	0·174 56	0·131 80	0·083 95	0·044 20	0·018 54	0·005 83	0·001 23	0·000 14	0·000 01	0·000 00	9
10	0·122 19	0·075 48	0·039 18	0·016 66	0·005 56	0·001 36	0·000 21	0·000 02	0·000 00	0·000 00	10
11	0·066 65	0·033 69	0·014 25	0·004 89	0·001 30	0·000 25	0·000 03	0·000 00	0·000 00	0·000 00	11
12	0·027 77	0·011 48	0·003 96	0·001 10	0·000 23	0·000 03	0·000 00	0·000 00	0·000 00	0·000 00	12
13	0·008 54	0·002 89	0·000 81	0·000 18	0·000 03	0·000 00	0·000 00	0·000 00	0·000 00	0·000 00	13
14	0·001 83	0·000 51	0·000 12	0·000 02	0·000 00	0·000 00	0·000 00	0·000 00	0·000 00	0·000 00	14
15	0·000 25	0·000 06	0·000 01	0·000 00	0·000 00	0·000 00	0·000 00	0·000 00	0·000 00	0·000 00	15
16	0·000 02	0·000 00	0·000 00	0·000 00	0·000 00	0·000 00	0·000 00	0·000 00	0·000 00	0·000 00	16

DIURNAL VARIA

Range of Diurnal Oscillation of the Thermometer.	JANUARY.		FEBRUARY.		MARCH.		APRIL.		MAY.		JUN
	No. of Variations.	Reduction to Unity.	No. of Variations.	Reduction to Unity.	No. of Variations.	Reduction to Unity.	No. of Variations.	Reduction to Unity.	No. of Variations.	Reduction to Unity.	No. of Variations.
0° to 1°	0	0·000	0	0·000	0	0·000	0	0·000	0	0·000	0
1 „ 2	8	0·026	1	0·003	1	0·003	1	0·003	0	0·000	1
2 „ 3	31	0·100	10	0·035	7	0·023	2	0·007	0	0·000	2
3 „ 4	61	0·197	27	0·096	27	0·087	5	0·017	3	0·010	0
4 „ 5	68	0·220	67	0·238	37	0·120	15	0·050	2	0·006	1
5 „ 6	50	0·162	69	0·245	38	0·123	26	0·087	12	0·039	8
6 „ 7	32	0·104	46	0·163	59	0·191	39	0·130	20	0·065	17
7 „ 8	22	0·071	35	0·124	52	0·168	51	0·170	23	0·075	30
8 „ 9	20	0·065	9	0·032	44	0·142	40	0·133	43	0·139	37
9 „ 10	8	0·026	10	0·035	19	0·061	38	0·127	43	0·139	44
10 „ 11	4	0·013	6	0·021	17	0·055	27	0·090	46	0·148	45
11 „ 12	3	0·010	2	0·007	6	0·019	21	0·070	49	0·158	42
12 „ 13	1	0·003	0	0·000	2	0·006	9	0·030	29	0·094	37
13 „ 14	1	0·003	0	0·000	0	0·000	8	0·027	21	0·069	16
14 „ 15	0	0·000	0	0·000	0	0·000	3	0·010	14	0·045	11
15 „ 16	0	0·000	0	0·000	0	0·000	4	0·013	2	0·006	4
16 „ 17	0	0·000	0	0·000	0	0·000	1	0·003	1	0·003	2
17 „ 18									2	0·006	2
18 „ 19											1
TOTAL ...	309		282		309		300		310		300

EMPERATURE.

JULY.	AUGUST.		SEPTEMBER.		OCTOBER.		NOVEMBER.		DECEMBER.	
Reduction to Unity.	No. of Variations.	Reduction to Unity.	No. of Variations.	Reduction to Unity.	No. of Variations.	Reduction to Unity.	No. of Variations.	Reduction to Unity.	No. of Variations.	Reduction to Unity.
0·000	0	0·000	0	0·000	0	0·000	0	0·000	0	0·000
0·000	0	0·000	0	0·000	1	0·003	9	0·030	11	0·036
0·000	1	0·003	0	0·000	9	0·029	16	0·053	34	0·110
0·003	0	0·000	3	0·010	15	0·048	40	0·134	55	0·178
0·023	2	0·006	12	0·040	47	0·152	62	0·207	67	0·217
0·032	8	0·026	27	0·090	56	0·181	61	0·204	65	0·210
0·078	22	0·072	42	0·140	55	0·178	48	0·160	32	0·104
0·110	37	0·119	51	0·170	34	0·110	31	0·104	27	0·087
0·133	50	0·161	59	0·197	36	0·116	20	0·067	12	0·039
0·136	47	0·151	42	0·140	23	0·074	4	0·013	4	0·013
0·133	50	0·161	27	0·090	17	0·055	7	0·023	1	0·003
0·123	37	0·119	18	0·060	7	0·023	1	0·003	0	0·000
0·084	25	0·082	11	0·037	8	0·026	0	0·000	1	0·003
0·061	16	0·052	5	0·017	1	0·003	0	0·000	0	0·000
0·053	9	0·029	2	0·007	0	0·000	0	0·000	0	0·000
0·019	5	0·016	0	0·000	0	0·000	0	0·000	0	0·000
0·010	1	0·003	1	0·003	0	0·000	0	0·000	0	0·000
0·003										
	310		300		309		299		309	

LETTER XXVI.

To complete as far as possible the Theory of Means, I give below three letters which M. Bravais, Professor of the Polytechnic School at Paris, has had the kindness to address to me, on the subject of a memoir *On the Valuation of Statistical Documents, and particularly on the Valuation of Means,* which I had inserted in the second volume of the *Bulletin of the Central Commission of Statistics of Belgium.* In this memoir I only had in view the examination of the case in which accidental causes have had no tendency to act in one direction more than another. Perhaps I had not sufficiently explained the object of that work. I the less regret it now, because it has given rise to very interesting letters on the part of a *savant* who has occupied himself with great success on the Theory of Probabilities, and its application to sciences of observation. I do not reproduce here the letters to which M. Bravais refers, because by doing so I should be only repeating in part the contents of many letters printed in that work.

" PARIS, 17*th January,* 1845.

" I have read with much interest your statistical researches. In them you make very intelligible to all the nature of the law of possibility of different kinds of results which a prolonged series of drawings may lead to. This law of possibility, when the number of drawings becomes very great, has for a limit e^{-hx^2}. But ought it to be admitted that, in nature, the curve of possibility of the deviations x, either more or less great, of a *physical* quantity, from its mean, shall always and *necessarily* be a function of this kind? I do not consider that this can be admitted *à priori,* although *à posteriori* observation mostly justifies this view. For example, in speaking of the height, may it not be conceived that special causes of preference exist which incline a certain number of human heights towards 1·700 metres, whilst other causes incline other heights towards 1·600 metres,—so that this curve of possibility presents two maxima, one towards 1·600, and the other to 1·700? Does the maximum possibility always correspond to the mean? Are the two branches of the curve always symmetrical? It appears to me that many examples might be cited where this is not so. As for instance, we know that the greatest deviations of the barometer towards the top of the column are scarcely half or two-thirds of the deviation of the barometer towards the base; so that we shall have a curve of possibility of the form

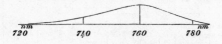

the two halves of which will not be symmetrical. But the mean ordinate will always divide the whole segment into two equal areas. May it not be conceived *à priori* that in the human height there is something analogous to that which presents itself in the barometrical column?

"I limit myself in pointing out my scruples to you," &c.

"PARIS, 27th January, 1845.

"I am really almost ashamed that a simple remark, which the perusal of your *Recherches Statistiques* suggested to me, should have inspired you with so much interest, and should have caused you in so flattering a manner to request my opinions on curves of possibility in general. And in the first place allow me to tell you that, having read again more carefully the 67th page of your work, I see clearly that you have in nowise fallen into the error which I intended to point out to you, since you say (Article I.) 'In the second case there exists in the results no necessary law of continuity.'

"As for the first part of your proposition, I am uncertain whether it is not enunciated in too general a manner, and if to be true it should not submit to important restrictions.

"Thus, and in the first place, it appears to me that, if the quantity to be measured (one well understood, and corresponding to a concrete really existing,) is susceptible of variation from unknown causes, the possibility ought not always to follow the same law. I will take, as you have done, the Polar Star as an example; only I will take a case rather more simple than yours; and I will suppose that it is required to obtain the true height of this star in its passage to the inferior or superior meridian, the element on which the declination of the star depends—the latitude—being known. To simplify still more, I will suppose that account has been kept of the precession, of the aberration, and of the nutation. Let us put aside the causes of error which may proceed from an inexact application of these corrections. There remains then several other causes of error which influence each isolated determination.* First, the parallax of declination (supposing it to be 1″). If the height obtained be 45° 30′, it will be seen that by this cause alone the height observed may vary between 45° 29′ 59″ and 45° 30′ 1″; but it is more easy to see that the heights will be oftener observed near these two extremes than near the mean value. In fact, the equation of the parallax will be of the form of $a \sin (a + b)$; a being the longitude of the sun, and a and b two constants. For simplification, let $a = 1$, $b = 0$; the equation reduces itself to $\sin a$, a quantity susceptible of all kinds of values between $+ 1$ and $- 1$. Calling e the error corresponding to a, $e + de$ that corresponding to $a + da$, we shall have $e = \sin a$, $de = \cos a \, da$. Let P be the

---

* The parallax of declination, the eccentricity in the repeating circle, the refraction, &c., would not be accidental errors, but should be treated as constant or variable errors, as M. Bravais has acknowledged in the following letter.

possibility that the error falls between 0 and e, and that the longitude of
the sun falls between 0 and a, we shall have $P = \phi\,(e)$, ϕ being an unknown
function.   We shall then have

$$\frac{d\,P}{da} = \frac{d\phi}{de}\,\frac{de}{da}.$$

And we have evidently

$$P = 2\,\frac{a}{2\pi}, \text{ hence } \frac{d\phi}{de} = \frac{1}{\pi\,cos\,a} = \frac{1}{\pi\,\sqrt{1 - e^2}}.$$

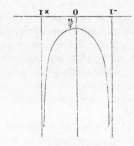

"It will be seen that the curve of the possibilities of the errors e is

$$p = \frac{1}{\pi\,\sqrt{1 - e^2}},$$

calling p the differential coefficient of P,—that is to say, the possibility of
the error e.   The curve has the form (as above) of a kind of parabola contained within the two asymptotes $e = 1$, $e = -1$.   Besides, if we integrate

$$\int \frac{1}{\pi\,\sqrt{1 - e^2}}\;de, \text{ from } -1 \text{ to } +1,$$

we find 1 for the integral, which conforms to that which should be obtained
à priori.   In this case the area contained between the curve and the axis
of the abscissa is finite, although the ordinates may be finite, for $e = \pm 1$.

"Here then is a first cause which tends to make the general curve of
possibility deviate from the exponential normal form e^{-hx^2}, and to give to it
two maxima corresponding to the two greatest parallactic digressions of the
star.

"I will now suppose that the repeating circle has been used, in which
the error of eccentricity $\frac{\varepsilon}{R}$ (ε being the eccentricity, and R the radius of the
circle,) is not sufficiently eliminated; the error of eccentricity will be on
each reading l, of the form $\frac{\varepsilon}{R}\,sin\,(l + a)$, a being a constant.   As there

are two readings l, l', of which the difference is twice the zenith distance (let it be $2Z$), the definite error is

$$\frac{\varepsilon}{R} \sin (l' + a) - \frac{\varepsilon}{R} \sin (l + a) = \frac{2\varepsilon}{R} \sin Z \sin [l + (z + a)].$$

Then, as Z, a, ε, R are constants, and as l may have all the possible values from $0°$ to $360°$, it will be seen that we fall again into the former case, with the exception of the value of the two constants, which will be generally different.

"Let us now consider the errors of refraction. We suppose refraction in its mean state, which corresponds to a value $0·08$ of the coefficient of the terrestrial refraction. Then, do we know the law of the differences of this coefficient above and below this mean value? This law depends on the variations of the decrease of the density in the lowest portions of the atmosphere. May not these variations be such that the law of possibility of the values of this coefficient about the mean $0·08$ differ from the general law e^{-hx^2}? This is what happens among others for the pressure of the air considered in its variation about its mean (about 760^{mm} from the level of the sea.) Is there a cause why the variations of the decrease of the densities should *à priori* be more regular and more symmetrical than those of the barometrical pressure? I do not think it can be affirmed. In the two cases there is an oscillation around a certain position of equilibrium.

"As for the errors of the graduation of the circle—errors which we should suppose variable if the circle is repeating,—(I will confine myself to consider this case, because in the non-repeating circle these errors are constant errors, which we are not to consider at present,)—instead of having for the error

$$e = \sin a, \text{ or } e = \frac{2\varepsilon}{R} \sin Z \cos (l + Z + a),$$

we have in this case $e = f(l)$; the function f being anything, consequently unassignable *à priori*. All the readings l being equally possible from 0 to 2π, the law of possibility will be of the form $\frac{1}{2\pi} \cdot \frac{dl}{de}$.

"The error of reading follows very different laws. It is very slight when the division of the vernier is precisely opposite that of the limb: it is greater when any division of the vernier is not opposite those of the limb. It depends on the remainder which the division of the reading l (by the arc which the vernier shows) leaves.

"The error of the parallax in the reading, as well as the error of coincidence of the star under the wire, proceed from causes still more difficult to analyze.

"It is the concurrence of all these causes which gives rise to definite error, and consequently to the law of possibility, which is discovered by observation *à posteriori*.

"It will come to about the same thing in the measures of the height of a single man taken at different times. The height is increased by repose, and diminished by fatigue. The variations are at least a third of an inch; but does the mean present itself most often ? We may doubt it.

"The moveable rule which serves for a measure may not always be horizontal. If the front and back portions of the vertical slide (which allows the moveable rule to pass up and down) are not quite parallel to each other, we shall have two different inclinations, acccrding to whichever of these two faces bears upon the corresponding face of the immoveable pillar, which serves as the measure; so that the result may be either a correction equal to $+ c$, or a correction equal to $- c$. The curve of possibility reduces itself to two points arranged symmetrically in relation to the axis of the ordinates of this curve.

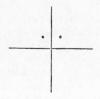

" There remains now the case of a quantity naturally perfectly invariable. I believe that even in this case the curve of possible errors may vary considerably, although these errors here all arise in the observation. But our instruments, and also our senses, carry in themselves causes of error. Thus, in the very simple case of a length which is measured by placing a compass alternately on the length to be measured and on a scale, there is a variation of the pressure of the hand on its branches, which may increase or diminish the result obtained. But is the cause corresponding to the pressure of the hand capable of producing an error from 0 to $\frac{1}{100}$ of a millimètre more probable than the cause capable of producing an error from $\frac{3}{100}$ to $\frac{4}{100}$ of a millimètre ? I do not think that we can affirm this *à priori*.

" In continuation, I think that generally every partial and distinct cause of error gives place to a curve of possibility of errors (or, if preferred, of differences about the mean), which may have any form whatever,—a curve which we may either be able or unable to discover, and which in the first case may be determined by considerations *à priori* on the peculiar nature of this cause, or may be determined *à posteriori* by observations on the isolated condition of other concomitant causes of error.

" I shall now endeavour to re-establish *in part* what I have just disproved, and to show the ordinary curve e^{-hx^2} following that which seems so often to agree with the observed differences about the mean result. I believe that three principal causes contribute to it.

" Firstly. The curve of possibility of the errors remains the same,—*a mean result* always approximates greatly to the form e^{-hx^2}, and the curve converges towards this form in the same ratio as the number of partial results concurring in the formation of the mean is augmented. It is in this, if I am not mistaken, that the law of great numbers consists. I am about to cite two examples of this convergence. Let us take again the case cited above, where the curve of possibility of error reduces itself to two points, the error being necessarily equal to $+ i$ or $— i$. Let us suppose the measure to be repeated, and let us see of what error the sum of the two measures will be susceptible: we shall have four combinations, which will correspond to $e = + 2i, 0, 0, — 2i$. If we have three measures, we shall have eight combinations, which I indicate by $+ + +, + + —, + — +,$ $+ — —, — + +, — + —, — — +, — — —$. We shall have four possible values, $e = + 3i, = + i, = — i, = — 3i$, with the probabilities $\frac{1}{8}, \frac{3}{8},$ $\frac{3}{8}, \frac{1}{8}$. In the general case of m measures, we shall have the possible errors

$$+ mi, + (m — 2) i \ldots — (m — 2) i — mi,$$

with the probabilities

$$\frac{1}{2^m}, m \frac{1}{2^m}, \frac{m (m — 1)}{1.2} \frac{1}{2^m} \ldots — m \frac{1}{2^m}, — \frac{1}{2^m}.$$

" We enter again upon the case which you have so well discussed in your memoir. The curve of possibility is discontinuous, and corresponds to a series of isolated points equidistant in the direction of the abscissæ, and the ordinates of which are proportional to the terms of the development of the binomial $(1 + 1)^m = 1 + \&c.$

" The continuous curve which unites these points to each other converges, as you have well demonstrated, towards the form $e^{— hx^2}$.

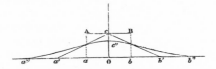

" The second example is that where the curve of possibilities is a right line AB, a finite right line parallel to the axis of the abscissæ, and cut into two equal parts in C by one of the ordinates; so that we have the possibility $p =$ a constant between two limits $\pm i$, and no part beyond these limits. On the figure we have $AC = CB = i$. Co should be equal to $\frac{1}{2i}$, so that the plane of the rectangle $aABb$ shall be equal to 1. This case presents itself when we take a number in a table; for example, a logarithm in a table of logarithms. Thus, in seeking log. 2, we find $0\cdot30103$: we can only affirm that log. 2 is comprised between $0\cdot301025$ and $0\cdot301035$, all intermediate values being equally possible. If we take the sum of two logarithms, what will be the curve of possibilities of errors to which this sum is subject? We find that this curve is represented by the bent line $a'Cb'$, $a'o$ being equal to $-2i$, and $b'o$ to $+2i$. For the sum of three logarithms, we have a system of three parabolas,—one diverging from a'' to the point of junction of the line Aa, and having for its equation

$$p = \frac{1}{6i} \cdot \frac{27}{2} \left(\frac{1}{2} + \frac{e}{6i} \right)^2;$$

the other diverging from the line Aa to the line Bb, and of which the equation is

$$p = \frac{1}{6i} \left[\frac{9}{4} - 27 \left(\frac{e}{6i} \right)^2 \right] \dots ;$$

and the third, like the first, which has terminated in b'' to a distance $ob'' = oa'' = 3i$, and has for its equation

$$p = \frac{1}{6i} \cdot \frac{27}{2} \left(\frac{1}{2} - \frac{e}{6i} \right)^2.$$

For four logarithms the curve would be composed of four parabolas of the third degree, and would begin greatly to recover the form of the curve e^{-hx^2}. Thus, in comparing it to the corresponding exponential, and having the same *maximum* ordinate, the probabilities corresponding to $e = 0$, $= \pm 2i = \pm 4i$ will be, for the exponential curve,

$$\frac{1}{8i}\ 2\cdot6667,\ \frac{1}{8i}\ 0\cdot6573,\ \frac{1}{8i}\ 0\cdot0098;$$

and for the curve of four parabolas

$$\frac{1}{8i}\ 2\cdot6667,\ \frac{1}{8i}\ 0\cdot6667,\ 0\cdot0000.$$

" The difference between the two curves of possibility is already of little importance.

" If, instead of the sums of m terms, we now consider the means of the same m terms, it is evident that the law of possibility of errors will be the same, replacing in the preceding results the error e on the sums, by the fraction $\frac{e}{m}$, which is the corresponding error on the means.

" There is, in Laplace's *Theory of Probabilities* (about pages 260 to 265), a general method of obtaining the law of possibility of error of a sum of numbers, when the error of each isolated number has a known law of possibility, which may be the same for all the numbers, or variable from one number to another. I have not at present this work by me, nor those of MM. Poisson and Cournot.

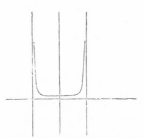

"But in every case, as the sum will be augmented, the curve will tend to approach the form e^{-hx^2}. It will be seen that the case least favourable to this rapid convergence is that in which the curve is enclosed by two vertical asymptotes ; but this case, carried to an extreme, brings us to a curve represented by two verticals, or (what comes to the same thing) represented by two points situated to the right and to the left of the axis of the ordinates. Thus this case, which is one already examined, is precisely that in which the convergence is the least rapid. Then the curve of the possibility of the error committed on the mean will always converge towards the form e^{-hx^2} as the number of the drawings increase, and that with a convergence superior to that of the series

$$1, \ m, \ \frac{m(m-1)}{2} \ \dots \ \frac{m(m-1)}{2}, \ m, \ 1$$

towards the form e^{-hx^2}.

"There is thus a first cause which tends often to bring back the curve of possibility e^{-hx^2} into the observation. This is the formation of the sums or of the means.

"Secondly. In the same case, where an isolated result is considered, a number proceeding from a single observation, the form e^{-hx^2} still tends to present itself, on account of the multiplicity of the *causes* of error of a *different nature* which intervene. The total error E is really the sum $e + e' + e'' + e''' \dots$ of a multitude of different errors which partly correct themselves, and to which the curves of possibility $\phi(e)$, $\psi(e')$, $\chi(e'') \dots$ &c. correspond, which tend to give a curve of possibility $\Phi(E)$ of errors E as much less different from the form e^{-hx^2} as the curves of possibility of partial errors are more numerous and less dissimilar to each other. The case in which all the curves would be the same enters into the case examined above. In proportion as the energy of these causes becomes on the contrary more and more dissimilar, the form of the function corresponding to the most energetic cause tends to predominate in the form of $\Phi(E)$.

"Thirdly. At length, I think in nature the curves of probabilities of differences are generally concave towards the axis of the abscissæ, which is as much as to say that the greatest differences are in general the most rare.

This is, I think, especially true for errors of observation. It seems *à priori* that this should be so; and what we know of the grouping of errors, or of differences where this grouping has been effectively observed, does not belie such a supposition.

"The question of the possibility of errors is extremely important in the sciences of observation, such as Astronomy, Geodesy, &c. I do not think it has yet been treated *ex professo* in a special work with desirable clearness and necessary developments. I have some notion that M. Hagen has tried to show that, in the measurement of an invariable natural magnitude, the possibility of error follows the exponential e^{-hx^2}; but, if I remember rightly, this demonstration did not appear sufficiently rigorous. In certain cases (for example, in Astronomy and Geodesy of Precision) each result is always deduced from a considerable number of measurements, and returning into the class of means its probable error should follow the law e^{-hx^2}.

"It is this special case that I have treated in a memoir communicated in 1837 to the Academy of Paris. This memoir being actually printed, I can soon send you a copy of it.

"I have frequently reflected upon this subject, but I have never had time to arrange my ideas. Thus I have availed myself with pleasure of the opportunity of submitting to you some of my ideas; this letter imposing upon me the necessity of giving them that form which hitherto they have not possessed. I do not know if I have been constantly perspicuous in the preceding exposition: it will be a pleasure to me to enter into any ulterior development which you may desire.

"Pardon this long letter, for the interest you take in these questions. When you do me the honour to write to me, be so good as to tell me whether you are of my opinion, or in what you differ from it. There is here a practical utility greater than is commonly imagined; it is in occupying myself with an entirely practical problem, of which the following is the enunciation: 'At what distance from the coast ought a ship to steer under sail, in order to take a survey of that coast?' (Well, this distance depends on the law of possibility of errors of the instruments, &c.) It is, I say, in considering this question that I have been led to investigate the probabilities in the natural sciences.

"Accept," &c.

"PARIS, 15th *February*, 1845.

"I am very far from regretting the publication of your memoir, which contains such excellent matter, and which, considered in a theoretical view alone, has the advantage of touching upon so many things which should, if possible, enter into the education of every man occupied in the study of nature, of whatever kind it may be. In France we are too ignorant of these principles. We have lately had here an attempt at the application of great numbers to medical statistics. The greater part of the medical profession have said much against it,—partly, doubtless, in consequence of a vague and instinctive fear of the danger of generalizations, and from a feeling of the want of the flexibility of therapeutics in following the special organization of disease; but also greatly from the difficulty of understanding the laws of means, and because the subject has not been treated with desirable perspicuity. I have only wished that in your memoir the restrictions which you have made, and which you have so well developed in your two letters, were put a little more in evidence; and I shall be far from being understood, if you have imagined my remarks intended anything else.

" I at first subscribed to the distinction to be made between variable and accidental causes; but I avow that this distinction appears to me difficult to establish fully, so that the accidental cause, attentively examined, should not pass entirely or partly into the rank of variable causes. What you say of causes which are regarded for a considerable time as constant, and then vary, appears to me to be very just; they are analogous to the secular equations of Astronomy; they are generally the most difficult to unravel, and often show their uncertainty in the final results.

"When the error to be feared in a result is to be determined, two things may be asked, either to determine an inferior or superior limit of this error. If, taking the first point of view, it is only concluded from these calculations that the degree of precision which may be accounted for is *at most* equal to the precision which the calculation assigns, I imagine that the physician is altogether right, and that he is not to be attacked on this ground. In the contrary case, where it is thought possible to place a superior limit to the possible error, it must be done, I imagine, with much prudence; for we have to fear too absolute an affirmation, and the constant errors as yet unknown, and all the *variable errors* for the long period of which you have spoken. But to place an inferior limit is already something. If this position had been taken for some time, many results would not have been too readily admitted to be afterwards overthrown. For example, the diminution of the horizontal magnetic intensity does not appear to me admissible as a demonstrated fact, although already we dare not to revoke it as a matter of doubt."

LETTER XXIX., page 137.

" The idea of filling up this gap in science has long made me sensible of the necessity of establishing as complete an enumeration as possible of periodical phenomena. I have thought it useful to submit it to the learned."

This is what I have done in many important cases, particularly at the 11th meeting of the British Association, at Plymouth, in July 1841. The following table of the principal phenomena to be observed has been inserted in the volume of their Proceedings for that session.

METEOROLOGY.

Pressure of the air during different months.

Temperature   ,,   ,,   ,,
Humidity   ,,   ,,   ,,
Electricity   ,,   ,,   ,,
Force and direction of the winds.
Quantity of rain, snow, &c.
State of the sky.
Meteors (shooting stars, auroræ borealis, &c.)

PHYSICS.

Magnetism of the earth.
Temperature of the earth at different depths.
Temperature of springs, of rivers, of the sea.
Radiation of heat.
Temperatures of vegetables and animals.
Phenomena of the seas.
Earthquakes.

CHEMISTRY.

Analysis of the air.
Analysis of rain-water.

BOTANY.

The foliation of plants and trees.
Blossoming   ,,   ,,
Fructification   ,,   ,,
Fall of leaves   ,,   ,,

AGRICULTURE.

Time of working the earth.
Time of ripening of vegetables.
Time of mowing.
Time of harvests.
Time of vintages.

ZOOLOGY.

Arrival, passage, and departure of migrating birds.
Arrival, passage, and departure of migrating fishes.
Appearance of different butterflies.
Entomological phenomena.
Reproduction in animals.
Mortality.

MAN.

Fecundation—Births.
Marriages.
Deaths and their nature—Suicides.
Diseases and their duration.
Mental alienation.
Crimes.
Consumption of fluids and food.
Movement of posts.
Movement of ports.
Movement of roads.

LETTER XXIX., page 139.

" With the study of the periodical phenomena of the atmosphere is connected that of the periodical phenomena relating to plants, animals, and man."

The introduction to the *Instructions pour l'Observation des Phénomènes Périodiques*, which I have published under the auspices of the Academy of Brussels, will give an idea of the plan which has been followed until now. It is this,—

"Whilst the earth travels in its annual orbit, it developes on its surface a series of phenomena which the periodical return of the seasons reproduce regularly in the same order. These phenomena, taken individually, have occu-

pied the attention of observers of all times; but they have generally neglected to study them collectively, and to seek for the laws of dependance
and of correlation which exist between them. The phases of the existence of
the least aphis of the most insignificant insect are allied to the phases of the
existence of the plant which nourishes it. This plant itself, in its successive
development, is in some manner the production of all the anterior modifications of the sun and atmosphere. This would be a very interesting study,
as it embraces at the same time all the periodical phenomena either *diurnal*
or *annual :* it would itself form a science as extended as instructive.

"It is, moreover, by the *simultaneousness* of observations made on a
great number of points that these researches may assume a high degree of
importance. A single plant studied with care would present to us most
interesting information. We might trace on the surface of the globe *syn-
chronic* lines for its foliation, its blossoming, its fructification, &c. The
lilac, for example, *Syringa vulgaris*, flowers in the environs of Brussels on
the 28th of April. We may conceive on the surface of the earth a line on
which the blossoming of this shrub takes place at the same time, as also
lines on which the blossoming is advanced or retarded ten, twenty, or
thirty days. Would these lines then be equidistant? Would they be
analogous to the *isothermal* lines? What would be the connection which
would exist between them? Again, would the *isanthesic* lines, or lines
of simultaneous blossoming, have a parallelism with the lines relating to
the foliation, or to other well-decided phases, in the development of the
individual plant? Let it be conceived, for example, that whilst the lilac
begins to flower at Brussels on the 28th of April, there are a series of places
to the north where this shrub is only putting forth its leaves. Does the line
which passes through these places correspond with the *isanthesic* line corresponding to the same time? It may still be asked whether those places
in which the foliation happens on the same day would also have the blossoming and fructification coincident. It will be already seen, in confining
one's self to very simple data, how many curious coincidences may be produced from a system of simultaneous observations established on a great
scale. The phenomena relating to the animal kingdom, particularly those
concerning the movements of migrating birds, would offer results no less
remarkable.

"Periodical phenomena can be divided into two great classes: one belongs
to the physical and natural sciences; the other is rather in the domain of
statistics, and concerns man living in the midst of the social state,—for even
society, with its tendency to withdraw itself as far as possible from the
natural laws, has not been able to escape from this periodicity which we
are now considering.

"*Natural* periodical phenomena are in general independent of *social* periodical phenomena; but it is not so with the latter in regard to the former.
Thus then a first step might be taken on this untrodden ground, and one
which seems to promise so much to the labours of those who know how to

work it; and thus may be commenced the simultaneous study of all the periodical phenomena which relate to the physical and natural sciences.

" These last phenomena themselves may be divided into several classes, and their study presupposes a considerable knowledge of meteorological phenomena on which they principally depend. Thus it is not unreasonable that Meteorology should take the initiative, and commence this series of continuous researches, to which observers who really aspire to follow Nature in all the laws of her organization and development ought henceforth to apply themselves.

" Nevertheless Meteorology, notwithstanding its persevering labours, has to the present time only been able to discover the mean state of the different scientific elements relating to the atmosphere, and the limits in which these elements may vary on account of climate and seasons. It must still continue its progress in conjunction with the study which we advocate; and to direct our judgment on observed results, it must show us at each step if the atmospheric influences are in a normal state, or if they manifest anomalies.

" The desire of applying myself to the study of periodical phenomena on a rather extended scale has led me to request many native and foreign scientific men to assist me with their views and observations. The favourable reception of my request has led me to believe that I have not deceived myself on the importance of the projected researches. I even see that it will be possible thus to compare our climate with that of neighbouring countries, and to obtain by direct and simultaneous observations (for Belgium in particular) valuable documents which we do not possess.

" Nevertheless, to proceed in a useful manner, it is necessary in the first place to observe according to the same plan; and it is not without reason that the scientific men whom I have addressed have generally asked for instructions concerning the objects to be observed, and the method to be followed in the observations, so as to render them *comparable*,—an essential quality for the desired end."

These instructions have been favourably received in various countries, and translated into different languages. I will particularly mention the translations into Russian by M. Kupffer, into German by MM. Mahlmann and Ch. Ritter in the Bulletin of the Geographical Society of Berlin, and by the Botanical Society of Ratisbon in its journal.

I take with pleasure the opportunity which here presents itself of thanking the numerous scientific men who have been willing to aid me in this *scientific crusade*. They are—

Belgium—MM. Blancquart, Cantraine, De Broe, De Selyslongchamps, Deville, Donckelaer, Forster, Galeotti, Kickx, Mac-Leod, Martens, Ch. Morren, Nève père, Robyns, Schwann, Spring, Spae, Van Beneden, Vincent, Wesmael, &c.

Netherlands—MM. Bergsma, Brants, Breitenstein, Martini Van Geffen, Reinwardt, Le Baron Thoe Schwartzenberg, Staring, Van Hall, Van Rees, Vrolik père.

France—MM. Bravais, Decaisne, Benoist, Fleurot, Le Baron d'Hombres Firmas, Valz.

Switzerland—MM. Agassiz, Chavannes, Delpierre, Espérandieu, Elie Wartmann.

England—Sir Dugald Brisbane, Messrs. Broun, Birt, Blackwall, Jonathan Couch, Leonard Jenyns.

Germany—MM. Brennecke, Léopold de Buch, De Martius, Lommler Schmid.

Italy—MM. Achille Costa, Colla, Passerini, Rondani, Scherer, Zantedeschi.

LETTER XXIX., page 140.

"The mortality is subject, at an interval of about six months, to a *maximum* and a *minimum.*"

The maximum happens in January, the minimum in July. Between these terms the other numbers regularly increase and decrease. This may be seen by the following table, which shows the mortality in Belgium during different months of the year, distinguishing between the inhabitants of towns and country.

MONTHS.	1815 to 1826		1841		1842		1843	
	Town.	Country.	Town.	Country.	Town.	Country.	Town.	Country.
January .	1·16	1·21	1·18	1·30	1·26	1·41	1·07	1·12
February .	1·09	1·20	1·19	1·32	1·20	1·31	1·10	1·25
March. .	1·05	1·19	1·10	1·15	1·10	1·17	1·17	1·32
April . .	1·00	1·12	1·12	1·17	1·12	1·17	1·05	1·17
May . .	0·95	0·98	0·95	0·98	1·00	0·95	1·01	1·05
June . .	0·90	0·88	0·92	0·89	0·95	0·87	0·91	0·95
July . .	0·87	0·81	0·84	0·79	0·83	0·77	0·88	0·83
August .	0·91	0·82	0·83	0·76	0·92	0·80	0·91	0·74
September	0·97	0·89	0·97	0·84	0·94	0·92	0·97	0·83
October .	1·00	0·93	0·93	0·88	0·87	0·86	0·96	0·89
November	1·02	0·94	0·98	0·94	0·89	0·88	0·96	0·91
December	1·08	1·03	1·00	0·99	0·92	0·89	1·01	0·94

LETTER XXXIII., page 163.

Calling the temperatures observed each day t, t_1, t_2, &c., we shall have

$$\Sigma t_n$$

to express the sum of the temperatures which produce the blossoming of a plant on the nth day from the time which serves for the point of starting, and which I call the instant of waking. According to my mode of calculation, we shall have

$$\Sigma t_n^2,$$

or rather

$$\Sigma t_n^2 + C.$$

This constant C depends on the state of the plant at the time of starting, and consequently on the winter and past season which have preceded it. This is still a consideration which Cotte thought might be neglected, and which I think must be attended to.

From the mode of calculation which I propose results the following corollary, which may give rise to interesting observations.

Variations of temperature are, all other things being equal, more favourable to the acceleration of vegetation than an uniform temperature. To explain,—if it is true that it is necessary to mark the temperatures in taking their second power, and not their first, it is evident that an individual mean temperature will be more or less efficacious according as it results from greater or less thermometrical variations. It is supposed, nevertheless, that these variations of temperature ought not to be carried beyond certain limits, which probably vary with the more or less delicate structure of the plants. Experience teaches us, so to speak, nothing on this interesting subject.

However, M. le Baron de Humboldt, whose works have contributed so much to extend the circle of our knowledge concerning the study of our globe, has been pleased to afford me several examples in support of this corollary, and particularly the manner in which the magnificent orange-trees at Berlin belonging to the King of Prussia are cultivated. I avail myself of the authority of this illustrious philosopher, since he has authorized me to do so.

I will here give the demonstration of the principle above enumerated.

Let T be the mean of the diurnal temperatures t, t', t'', &c., and let Δ, Δ', Δ'', &c., $- \Delta''' - \Delta^{\text{IV}}$, — &c. be the variations or differences of the temperatures t, t', t'', &c., compared with this mean T: we shall have

$$t^2 = (T + \Delta)^2 = T^2 + 2\Delta T + \Delta^2,$$

$$t'^2 = (T + \Delta')^2 = T^2 + 2\Delta'T + \Delta'^2,$$

$$\cdot \qquad \qquad \cdot$$
$$\cdot \qquad \qquad \cdot$$
$$\cdot \qquad \qquad \cdot$$

$$t'''^2 = (T - \Delta''')^2 = T^2 - 2\Delta'''T + \Delta'''^2 ;$$

$$\cdot \qquad \qquad \cdot$$
$$\cdot \qquad \qquad \cdot$$

whence

$$\Sigma t^2 = nT^2 + 2T (\Delta + \Delta' + \Delta'' + \&c. - \Delta''' - \Delta^{IV} - \&c.) + \Sigma\Delta^2.$$

But, according to the property of the mean, we shall have

$$\Delta + \Delta' + \Delta'' + \&c. - \Delta''' - \Delta^{IV} - \&c. = 0,$$

which reduces the preceding equation to

$$\Sigma t^2 = nT^2 + \Sigma\Delta^2.$$

Preserving for T the same value, the last quantity will become a minimum when $\Sigma\Delta^2$ is nothing,—that is to say, when there is no longer any variation. Thus the sums of the squares of the temperatures Σt^2 will be the smallest possible, the mean temperature T of the period remaining the same when the variations Δ are nothing.

LETTER XXXV., page 182.

" Collecting statistics is generally well understood. But it is not so with the definition of this science."

It cannot enter into my views to make known here all the definitions which have been given of statistics, still less to discuss their value. I think, nevertheless, that I ought to remark that statists have generally differed upon an essential point. Some reduce everything to numbers, and make the science consist of a vast series of tables; others, on the contrary, seem to fear numbers, and look upon them as giving only superficial and incomplete ideas of facts. These two extremes are equally injurious; but it may be truly said, that when results can be estimated by figures, it is always to figures that we must recur to obtain exact estimates and comparable results.

Others, and particularly the learned economist J. B. Say, would exclude from statistics everything which is not essentially variable. "It is the object of this science," he says, "to prove the real state of things whose conditions may successively change, and not an immutable state of things." This definition, although very ingenious, has often been objected to as being too limited.

My learned friend the Dr. Villermé, who has enriched statistics with many works, has given a definition of this science which appears to me to be very able. I should only have wished that he had not excluded the idea of the time, which appears essential. This is how Dr. Villermé expressed himself in opening a course of statistics at the Athénée Royal de Paris. "Statistics is the exposition of the state, of the situation, or (as Achenwall says) of everything which produces effect in political society, in a country, in a certain place. But it is agreed that this exposition, freed from interpretations, from theoretic views, from all system, and consisting (so to speak) of a simple inventory, may be digested in such a manner that all results may be easily compared, that one may easily be brought to bear upon another, that their mutual dependance and the general effect of institutions may be perceived—the happiness or unhappiness of the inhabitants, their prosperity or their adversity, the strength or the weakness of the people, may be deduced from them."

Schlözer of Göttingen, comparing the statistics of history, endeavours to point out the line of demarcation which the consideration of time establishes between the two sciences. "History," says he, "is statistics in movement, and statistics is history in repose,—*Geschichte ist eine fortlaufende Statistik, und Statistik eine stillstehende Geschichte.*" And he adds that history is the whole, and statistics is a part of it.

The learned geographer M. Adrien Balbi has endeavoured to show on his part the distinction which he establishes between statistics and geography.

M. Théodore Fix has lately published, in the *Journal des Economistes* of September 1845, an article entitled *De la manière d'observer les Faits Economiques*, which contains very enlightened views on the general defects of statistics. The author may perhaps be considered to entertain a very decided contempt for statistics, when he himself makes use of the following: "We object only to the defects of statistics, to the intemperance and the negligence with which numbers are accumulated, and to the pretension of some men who consider themselves economists only because they have collected facts and arranged numbers with more or less discernment."

Letter XXXVII., page 191.

"Some writers have further distinguished the sources, according to the degree of confidence they merit, into *primary* and *secondary* sources."

Professor Mone, in his *Theory of Statistics*, moreover establishes the following subdivisions :—

Primary sources,—

    I. Charters and treaties.

    II. Official publications.

Secondary sources,—

   I. Works published by native authors.
   II. Accounts of travels written by foreigners.
   III. Daily journals and periodical works.
   IV. Correspondence.

M. J. Fallati, in his work *Einleitung in die Wissenschaft der Statistik*, Tubingue, 1843, divides statistics into concrete, abstract, and assumed. The same author then admits a great number of subdivisions, in order that statistical researches should apply themselves to parts more or less complete of the social organization, and to more or less extended portions of the universe.

M. Dufau, in his *Traité de Statistique*, Paris, 1840, divides statistics into general, particular, local, and special. " We say that statistics is *general* when it treats of objects of universal nature and comprehends all countries. . . . . We call it *particular* when it only relates to a particular country, as France or England; *local*, when the facts it embraces only concern a town or territorial boundary, such as a province or a department. Lastly, statistics take the name of *special* when applied exclusively to one class of facts. Even the very nature of the objects which it is to determine in this latter case following the divisions and subdivisions of *physical, meteorological, medical*, and such like statistics." (Page 83.)

Letter XLIII., page 227.

" Constant ratios are established between these three things,—the general number of crimes committed, the number of crimes known, and the number of crimes prosecuted."

Let us suppose that C represents all the crimes committed in a province, those which have been prosecuted as well as those which have not, and that c represents only those which have been prosecuted. The ratio $\frac{c}{C}$ will give the amount of crime repressed in that province; it will indicate how many crimes have been prosecuted out of the number committed and brought to the knowledge of justice. If we have reason to believe that the repression $\frac{c'}{C'}$ is the same in another province, we shall have

$$\frac{c}{C} = \frac{c'}{C'}, \text{ whence } \frac{c}{c'} = \frac{C}{C'};$$

that is to say, that there exists between the crimes committed and prosecuted of both provinces the same relation as between all crimes in general.

The very judicious reflections inserted by M. Alphonse Decandolle in the *Bibliothèque Universelle de Genève*, April 1830, article *Revue des Progrès de la Statistique*, may be consulted on this subject.

Letter XLIV., page 235.

" I extract them from a small work which I have just received," &c.

Noticing in the *Revue Médicale de Paris,* in the number for November 1840, a work of M. Gavarret entitled *Principes généraux de Statistique Médicale,* M. le Dr. Martins has well explained the use which may be made of statistics in the medical sciences, and its insufficiency on many occasions. It is evident that the greater part of the very animated discussions which have arisen on this subject, in almost all medical academies and societies, rest upon the mistakes or the excessive pretension of some statists. MM. Villermé, Benoiston, De Châteauneuf, Louis, Andral, Magendie, Civiale, Mélier, Lévy, Lélut, Leuret, Mitivié, Falret, Bazin, Lombard, Mallet, D'Espine, Parchappe, &c., have proved, in France and at Geneva, the advantages which may be derived in medicine from well-directed statistical studies ; in Germany, MM. Burdach, Casper, Carus, Tiedeman, Riecke, Friedlander, Heyfelder, Valentin, Schwann, Gluge, &c. England (always so quick to perceive the useful side of things), Italy, and most other civilized countries, may equally number many learned physicians who have made a good use of statistics.

Letter XLVI., page 246.

" Belgium and Sardinia have first entered upon this path."

It will doubtless be pleasing to find here an exposition of the motives which led to the formation of the Belgian Central Commission of Statistics.

" Sire,

" In creating in the Ministry of the Interior a central office of statistics, the Provisional Government proposed to employ its administration both in collecting and classing in methodical order accurate and complete documents on all points which should be the object of this important branch of the science of government.

" But by degrees this end has been departed from. Many departments entirely neglected statistics; others worked at them partially, drawing sometimes from the same sources, meeting and crossing one another in their researches. This want of unity could but inevitably lead to discrepancies, double entries, and omissions. Nevertheless, the partial publications emanating from the different departments possessed a real merit. The accounts rendered (so interesting in a moral point of view) of criminal justice, the territorial statistics, the tables of commerce, the general documents, of which five volumes have already appeared, are works of great importance, and which attest at each period a new progress of the administration.

"But what our statistics need, so that science and the government may derive from them all the fruits which they have a right to expect after so many efforts, are a single direction, a precise end, and a perfectly determined basis of investigation. The result of the measure which I have just proposed to your Majesty would be, in the first place, to ensure these essential qualities of statistics.

"It would create a Central Commission of Statistics.

"Each department would be there represented by one or more delegates, chosen by the minister from the *employés* who have made a special study of statistics, and examined thoroughly into the branches connected with his department. The assembly of these delegates would be presided over by a man of science, versed in social economy, and accustomed to sum up statistical works.

"The nature of the task of the Central Commission of Statistics is easily deduced from what I have said above of the defects of the actual system. To make all the scattered documents which the different administrations now collect converge towards a common centre should be its aim.

"Thus it will point out the gaps and the superfluous details of actual publications.

"It will propose models of systems and tables for gathering together and classifying the elements of these publications.

"It will take care that all duplicate labour should be avoided in requiring information, and in the publications themselves.

"It will correspond directly with the Minister of the Interior: it will submit to him its observations and propositions, with the necessary instructions for each department. The Minister of the Interior will communicate the views of the commission to his colleagues, who will be at liberty to adopt or to modify them.

"Each department will continue to publish the statistics which concerns itself; but a uniform plan having been previously adopted, unity and uniformity will be substituted for the diversity of the actual publications.

"It is with the unanimous consent of my colleagues, Sire, that I submit this project for your Majesty's sanction.

"If, as it may be hoped, the commission worthily fulfils the end which we propose in instituting it, government, the legislative chambers, the country, will find in the official statistical publications authentic facts calculated to enlighten all discussions, to urge on useful works, and to show each year the situation, the strength, and the material and moral resources of the kingdom.

"*The Minister of the Interior,*

"LIEDTS."

The royal decree establishing the commission was couched in these terms:—

"LEOPOLD, KING OF THE BELGIANS.

"To all present and to come, greeting.

"Considering the decree of the Provisional Government of Belgium, dated 24th January 1841, charging the Minister of the Interior with the preparation of a general statistical account of the kingdom;

"Wishing to regulate and extend the statistical publications of the different ministerial departments;

"On the report of our Minister of the Interior, and under the advice of the other chiefs of the departments,

"We have decreed and do decree,—

"Art. 1. There is instituted under the Minister of the Interior a Central Statistical Commission, whose members, selected as far as possible from among the functionaries of the different ministers, shall be nominated by us.

"Art. 2. One-third part of the commission shall be renewed every second year, from the 1st January 1843. The quitting of office shall take place in order of seniority, or by lot in cases of equal seniority.

"The members quitting office may be retained.

"Art. 3. The commission shall propose a complete plan for the publication of statistical documents in reference to the different branches of the administration.

"Art. 4. It shall advise on all communications which shall be addressed to it by our Minister of the Interior.

"It shall correspond directly with this minister.

"Art. 5. The mode of exercising its powers and the order of its works shall be determined by a special regulation, drawn up by the Minister of the Interior, in concert with the chiefs of other departments, and shall be submitted for our approbation.

"Art. 6. It shall be allowed a sum for personal attendance and office expenses.

"Art. 7. Our Minister of the Interior is charged with the execution of the present decree.

"Given at Brussels, this 16th March 1841.

"LEOPOLD."

Sometime afterwards the following decree was issued, on the proposition of M. Nothomb, then Minister of the Interior, and completed the statistical organization of the kingdom.

" LEOPOLD, KING OF THE BELGIANS.

" To all present and to come, greeting.

" With reference to Article 3 of our decree of the 20th October 1841, thus worded,—

"§ 1st. 'Provincial or local statistical commissions may be established.'

" In modification of the second paragraph of the said decree, thus worded,—

"§ 2nd. 'The members of these commissions shall be nominated by the Minister of the Interior, on the proposition of the Central Commission.'

" On the proposal of our Minister of the Interior,

" We have decreed and do decree,—

" Art. 1. There is established at the chief place of each of the provinces of the kingdom a commission charged with co-operating in the labours of the Central Statistical Commission.

" Art. 2. One-third part of each provincial statistical commission shall be renewed every second year, from 1st January 1845. The quitting office shall take place in order of seniority, or by lot in cases of equal seniority. The retiring members may be retained.

" Each commission shall consist of not more than twelve, nor less than six members, exclusive of the president.

" Art. 3. The governor is by right the president of the statistical commission for his province. He may appoint as his substitute a member of the permanent deputation of provincial council.

" The commission names a vice-president, in case of the absence of the president or his delegate.

" It selects its secretary from among its members.

" Art. 4. The commissions superintend and regulate the statistical labours in the provinces ; they cause to be collected the returns required of them, or which they may consider it useful to collect, and give their advice upon documents which are officially transmitted to them.

" Art. 5. The governors appoint, when necessary, the *employés* of the Provincial Government to work under the superintendence and control of the commissions. These *employés* remain under the authority of the governors, who, in order to ensure the execution of the work, alone give the orders they may think right.

" Art 6. The governors place at the disposition of the commissions a place wherein to hold their sittings, and office furniture.

" There may be allowed to them under this head, concurrently with the sums appropriated to statistical works in the provincial budgets, a subsidy chargeable on the credit carried to the budget of the department of the Interior for the expense of publishing of general statistics.

" Art. 7. Within three months of the installation of the commissions, they shall address a project for regulating their plan and powers to the Minister of the Interior, who shall decree it after having heard the Central Commission.

" ART. 8. Our Minister of the Interior is charged with the execution of this decree.

" Given at Brussels, the 6th April 1845.

" LEOPOLD."

The commission has up to the present time published, under the title of *Bulletin*, two volumes in quarto of its memoirs and *procès-verbaux;* it has, in addition, ordered the publication of five volumes on the population and the movement of the civil state, the production of which is entrusted to M. Heuschling, the secretary of the commission.

LETTER XLVI., page 246.

Many states have well-regulated statistical departments directed by able men. I would mention above all that of Berlin, which takes the first rank (thanks to the talent of MM. Hoffmann and Dietrici) ; that of London, of which Mr. Porter is the head ; those of Bavaria, of Wurtemburg, of the Grand Duchy of Baden, &c. In the different administrations of France, we also find learned and zealous men charged with the publication of official documents; and we have but to regret that there exists no tie amongst them.

Different societies are also organized for the development of statistical studies. That of London includes a great number of distinguished men, who have contributed, by their writings and the publication of a journal, to give a powerful impulse to this science. Similar societies have been established at Glasgow, Manchester, Paris, Frankfort, Lubeck, in Prussia, in Saxony, in the Grand Duchy of Hesse-Darmstadt, &c.

Statistics has equally received assistance from the establishment of classes for the teaching of the science, especially in Germany, where it has for its organs MM. Schubert, Schnabel, Schoen, R. Von Mohl, F. B. Weber, Voigtel, L. Moser, Rau, Bernouilli, Fraenzl, &c. In France the science is taught at the Conservatoire des Arts et Metiers, by M. le Baron Charles Dupin, whose works have contributed so much towards making known the agriculture and manufactures of France. M. Ramond de la Sagra has, for his part, endeavoured to render it popular in Spain.

LETTER XLV., page 238.

" There is then a *maximum* which may be obtained, but which can only be determined by the aid of good statistical documents."

Let us call x the amount of fare, and y the number of passengers corresponding to the fixing of this fare. There will be a relation between x and y,

$$y = f(x).$$

But if the tariff of fares has changed several times, and all other circumstances have remained the same, we shall have the co-ordinates for the points of the curve represented by the preceding equation. These points may aid in constructing and calculating empirically the curve, and consequently in deducing the *maxima* and *minima* values of the ordinate.

The most simple hypothesis that can be made is to suppose that the number of passengers increase in proportion as the fares diminish, or rather that it is in the inverse ratio of the fares. In this case the equation will be

$$y = \frac{a}{x}, \text{ whence } yx = a,$$

which indicates an equilateral hyberbola referred to its asymptotes. If $x = 0$, $y = \infty$,—that is to say, supposing the fare to be nothing, the number of travellers will be infinitely great. On the other hand, making $x = \infty$, $y = 0$,—that is to say, supposing the fare to be infinitely great, the number of travellers will be nothing. We may conceive that these mathematical hypotheses will never be realized, and that observations will never extend to these extreme limits: we may also conceive that, in most cases, the curve will not differ much from an equilateral hyberbola.

It would be desirable that geometrical constructions should be employed to represent the influences which prices may exert in certain social elements to which they are related. Great advantage might be drawn from them.

C. & E. Layton, Printers, 150 Fleet Street, London.

THE DEVELOPMENT OF SCIENCE

An Arno Press Collection

Electro-Magnetism. 1981

Gravitation, Heat, and X-Rays. 1981

Laws of Gases. 1981

Theory of Solutions and Stereo-Chemistry. 1981

The Wave Theory of Light and Spectra. 1981

Ackerknecht, Erwin H. Rudolf Virchow. 1953 *and* Schwalbe, J., editor. Virchow-Bibliographie, 1843-1901. 1901

Airy, George Biddell. Gravitation. 1834

Anderson, David L. The Discovery of the Electron. 1964

Beer, John. The Emergence of the German Dye Industry. 1959

Brown, Theodore. The Mechanical Philosophy and the "Animal Oeconomy." 1981

Candolle, Alphonse de. Histoire des sciences et des savants depuis deux siècles. 1885

Cheyne, Charles H.H. An Elementary Treatise in the Planetary Theory. 1883

Cohen, I. Bernard, editor. Andrew N. Meldrum. 1981

Cohen, I. Bernard, editor. The Conservation of Energy and the Principle of Least Action. 1981

Cohen, I. Bernard, editor. The Leibniz-Clarke Correspondence. 1981

Cohen, I. Bernard, editor. Studies on William Harvey. 1981

Coleman, William, editor Carl Ernst Von Baer on the Study of Man and Nature. 1981

Coleman, William, editor. French Views of German Science. 1981

Coleman, William, editor. Physiological Programmatics of the Nineteenth Century. 1981

Domson, Charles. Nicolas Fatio de Duillier and the Prophets of England. 1981

Donahue, William H. The Dissolution of the Celestial Spheres. 1981

Farrell, Maureen. William Whiston. 1981

Gardner, Walter M., editor. **The British Coal-Tar Industry.** 1915

Godfray, Hugh. **An Elementary Treatise on the Lunar Theory.** 1871

Graetzer, Hans G. and David L. Anderson. **The Discovery of Nuclear Fission.** 1971

Grimaux, Édouard. **Lavoisier: 1743-1794.** 1888

Hall, Diana Long. **Why Do Animals Breathe?** 1981

Hall, Maria Boas. **The Mechanical Philosophy.** 1981

Hannequin, Arthur. **Essai critique sur l'hypothèse des atomes dans la science contemporaine.** 1899

Harvey-Gibson, Robert J. **Outlines of the History of Botany.** 1919.

Heidel, William Arthur. **Hippocratic Medicine.** 1941

Heilbron, John L. **Historical Studies in the Theory of Atomic Structure.** 1981

Helm, Georg. **Die energetik.** 1898

Herschel, J.F.W. **Essays from the Edinburgh and Quarterly Reviews.** 1857

Hiebert, Erwin N. **Historical Roots of the Principle of Conservation of Energy.** 1962

Hilts, Victor L. **Statist and Statistician.** 1981

Hirschfield, John Milton. **The Academie Royale des Sciences (1666-1683).** 1981

Home, Roderick Weir. **The Effluvial Theory of Electricity.** 1981

Kendall, Maurice G. and Alison Doig. **Bibliography of Statistical Literature.** Three volumes. 1962, 1965 and 1968

Maier, Clifford L. **The Role of Spectroscopy in the Acceptance of the Internally Structured Atom, 1860-1920.** 1981

Meyer, Kirstine. **Die entwickelung des temperaturbegriffs im laufe der zeiten.** 1913

Milne-Edwards, Henri. **Introduction à la zoologie générale.** 1853

Morgan, Augustus de. **An Essay on Probabilities.** 1838

Mouy, Paul. **Le développement de la physique cartésienne 1646-1712.** 1934

Olmsted, J.M.D. **Francois Magendie.** 1944

Partington, J.R. and D. McKie. **Historical Studies on the Phlogiston Theory.** 1937, 1938 and 1939

Petit, Gabriel and Maurice Leudet. **Les allemands et la science.** 1916

Priestley, Joseph. **History and Present State of Discoveries Relating to Vision, Light, and Colours.** 1772

Quetelet, M.A. **Letters Addressed to H.R.H. The Grand Duke of Saxe Coburg and Gotha, on the Theory of Probabilities, as Applied to the Moral and Political Sciences.** 1849

Roe, Shirley A., editor. **The Natural Philosophy of Albrecht von Haller.** 1981

Sayili, Aydin. **The Observatory in Islam.** 1960

Schofield, Christine Jones. **Tychonic and Semi-Tychonic World Systems.** 1981

Schweber, S.S., editor. **Aspects of the Life and Thought of Sir John Frederick Herschel.** 1981

Shirley., John W., editor. **A Source Book for the Study of Thomas Harriot.** 1981

Struve, Friedrich George Wilhelm. **Études d'astronomie stellaire.** 1847

Turner, Dorothy Mabel. **History of Science Teaching in England.** 1927

Woolf, Harry. **The Transits of Venus.** 1959

Wurtz, Adolf. **A History of Chemical Theory, from the Age of Lavoisier to the Present Time.** 1869

Youmans, Edward L., editor. **The Correlation and Conservation of Forces.** 1865

Zloczower, A. **Career Opportunities and the Growth of Scientific Discovery in Nineteenth Century Germany.** 1981